Evaluating What Works

Those who work in allied health professions and education aim to make people's lives better. Often, however, it is hard to know how effective this work has been: would change have occurred if there was no intervention? Is it possible we are doing more harm than good? To answer these questions and develop a body of knowledge about what works, we need to evaluate interventions. Objective intervention research is vital to improve outcomes, but this is a complex area, where it is all too easy to misinterpret evidence. This book uses practical examples to increase awareness of the numerous sources of bias that can lead to mistaken conclusions when evaluating interventions. The focus is on quantitative research methods, and exploration of the reasons why those both receiving and implementing intervention behave in the ways they do. *Evaluating What Works: An Intuitive Guide to Intervention Research for Practitioners* illustrates how different research designs can overcome these issues, and points the reader to sources with more in-depth information. This book is intended for those with little or no background in statistics, to give them the confidence to approach statistics in published literature with a more critical eye, recognise when more specialist advice is needed, and give them the ability to communicate more effectively with statisticians.

Key Features:
- Strong focus on quantitative research methods
- Complements more technical introductions to statistics
- Provides a good explanation of how quantitative studies are designed, and what biases and pitfalls they can involve.

Dorothy Bishop was Professor of Developmental Neuropsychology at the University of Oxford from 1998 to 2022. Dorothy is a Fellow of the Academy of Medical Sciences, a Fellow of the British Academy, and a Fellow of the Royal Society. She been recognised with Honorary Fellowships from the Royal College of Speech and Language Therapists, the British Psychological Society, and the Royal College of Pediatrics and Child Health. She has Honorary Doctorates from the Universities of Newcastle upon Tyne, UK, Western Australia, Lund, Sweden, École Normale Supérieure, Paris, and Liège, Belgium. She is an Honorary Fellow of St John's College, Oxford.

Paul Thompson is an Assistant Professor in Applied Statistics and the department lead for statistics and quantitative methods at the Centre for Educational Development, Appraisal and Research (CEDAR) at the University of Warwick. Between 2014 and 2021 he worked at Oxford University within the Department of Experimental Psychology, working on a wide range of projects including behavioural, genetics, and neuroimaging (brain scanning) studies in developmental language disorders such as Dyslexia, and Developmental Language Disorder, and language development in those with learning and developmental disabilities, such as Down Syndrome and Autism.

Evaluating What Works
An Intuitive Guide to Intervention
Research for Practitioners

Dorothy V. M. Bishop and Paul A. Thompson

CRC Press
Taylor & Francis Group
Boca Raton London New York

CRC Press is an imprint of the
Taylor & Francis Group, an **informa** business

A CHAPMAN & HALL BOOK

Designed cover image: © Shutterstock ID 2281563993, Photo Contributor PeopleImages.com - Yuri A

First edition published 2024
by CRC Press
2385 NW Executive Center Drive, Suite 320, Boca Raton FL 33431

and by CRC Press
4 Park Square, Milton Park, Abingdon, Oxon, OX14 4RN

CRC Press is an imprint of Taylor & Francis Group, LLC

© 2024 Dorothy V. M. Bishop and Paul A. Thompson

ISBN: 978-1-032-59120-9 (hbk)
ISBN: 978-1-032-59061-5 (pbk)
ISBN: 978-1-003-45307-9 (ebk)

DOI: 10.1201/9781003453079

Typeset in Latin Modern font
by KnowledgeWorks Global Ltd.

Publisher's note: This book has been prepared from camera-ready copy provided by the authors.

Contents

Preface

Why did we write this book?

Those who work in allied health professions and education aim to make people's lives better. Often, however, it is hard to know how effective we have been: would change have occurred if we hadn't intervened? Is it possible we are doing more harm than good? To answer these questions and develop a body of knowledge about what works, we need to evaluate interventions.

As we shall see, demonstrating that an intervention has an impact is much harder than it appears at first sight. There are all kinds of issues that can arise to mislead us into thinking that we have an effective treatment when this is not the case. On the other hand, if a study is poorly designed, we may end up thinking an intervention is ineffective when in fact it is beneficial. Much of the attention of methodologists has focused on how to recognize and control for unwanted factors that can affect outcomes of interest. But psychology is also important: it tells us that our own human biases can be just as important in leading us astray. Good, objective intervention research is vital if we are to improve the outcomes of those we work with, but it is really difficult to do it well, and to do so we have to overcome our natural impulses to interpret evidence in biased ways.

Who is this book for?

The inspiration for the book came from interactions with speech and language therapists (known as speech-language pathologists or logopeds outside the UK), and the illustrative cases are mostly from that discipline, but the basic principles covered here are relevant for any field where a practitioner aims to influence outcomes of those they work with. This includes allied health professions and education.

In medicine, practitioners who wish to evaluate treatments typically have access to expert statisticians and methodologists, who can advise them on the most efficient research designs and up-to-date analyses. They may also be able to access substantial amounts of funding to cover the salaries of experienced trials personnel. However, in many fields, professionals who administer interventions have limited training in research design and little or no support from professional statisticians or methodologists. Training in most professions allied to health and education does not usually include much instruction in research methodology and statistics – particularly as this relates to evaluation of interventions, which is a complex and thorny topic. It seemed, therefore, that there was a need for a basic text that would explain the rationale and potential pitfalls of intervention research, as well as providing a template for good practice in the evaluation and design of intervention studies.

What is covered?

This book is not a how-to-do-it manual, so much as a why-to-do-it. Our main goal is to instill in the reader awareness of the numerous sources of bias that can lead to mistaken conclusions when evaluating interventions. Real-life examples are provided with the aim of providing an intuitive understanding of these issues. Of course, it is not much use telling people what *not* to do if you don't also give guidance on approaches that are effective. We will illustrate ways in which different research designs can overcome problems, but it is beyond the scope of this book to give detailed instructions on how to implement different methods: instead, we will point readers to other sources that give more in-depth information.

Our focus is on quantitative research methods. Qualitative research methods are increasingly recognized as providing an important complementary perspective on intervention research, by throwing light on the reasons why people – both those receiving intervention and those implementing it – behave as they do. Sometimes it makes more sense to do a qualitative study to refine a research question before diving in using methods that require that everything be converted into numbers (Greenhalgh & Taylor, 1997). If you feel that a quantitative study is missing out on something essential and important about how and why intervention works, this may be an indication that a qualitative study is needed to scope out the problem more fully. Our expertise, however, is with quantitative methods, and our aim is to write a basic explainer of how such studies are designed and what biases and pitfalls they can involve.

The intended readership is those who have little or no background in statistics. Lack of statistical training is a massive obstacle to practitioners who want to do intervention research: it not only makes design and analysis of a study daunting, but it also limits what messages the reader can take from the existing literature. This book should be seen as complementing rather than substituting for a technical introduction to statistics. Many readers may be reluctant to study statistics in more depth, but it is hoped that the account given here will give them confidence to approach statistics in the published literature with a more critical eye, to recognize when the advice of a professional statistician is needed, and to communicate more effectively with statisticians. We recommend Russell Poldrack's open source text, as a complement to this book for those who wish to learn more about statistics (2018).

Intervention research is a branch of science, and you can't do good science without adopting a critical perspective – on the research of yourself as well as others. We hope this book will make it easier to do that and so to improve intervention research in a range of non-medical fields.

Acknowledgements

We owe a huge debt of gratitude to Yihue Xie, whose work developing the *bookdown* package made this book possible. It is no exaggeration to say that we would not have attempted to write a book on this topic if we hadn't had *bookdown* to make it easy. The book ModernDive: Statistical Inference via Data Science (2023) was also highly influential in demonstrating what could be done with *bookdown*, and we frequently visited their code to check out ideas for formatting and packaging information.

Our thanks also to James Pustejovsky whose recommendations were helpful in relation to single case designs, and to Daniël Lakens, Daniel Sutherland, Nicola Dawson, Lucy Hughes, and an anonymous reviewer, who read the entire text and offered invaluable feedback.

Authors

Dorothy V. M. Bishop

Dorothy Bishop was Professor of Developmental Neuropsychology at the University of Oxford from 1998 to 2022. She was funded for 20 years by the Wellcome Trust on a Principal Research Fellowship heading up a programme of research into children's communication impairments. In 2017 she embarked on running an ERC Advanced Award on the Nature and Origins of Cerebral Lateralisation. She retired in 2022, but continues with some academic activities, particularly in the sphere of open and reproducible research.

Dorothy's interest in cognitive disorders was stimulated when she studied Experimental Psychology at Oxford University in the early 1970s, and she went on to train as a clinical psychologist at the Institute of Psychiatry, in the days when this involved a two-year M.Phil. She particularly enjoyed neuropsychology and welcomed an opportunity to return to Oxford to work at a Neuropsychology Unit in the Radcliffe Infirmary where Freda Newcombe was her supervisor. Freda steered Dorothy's interest toward the numerous fascinating cases who were referred from the Park Hospital for Children, and this launched her career as a specialist in developmental language disorders.

She was fortunate in receiving long-term research funding, first from the Medical Research Council and subsequently from the Wellcome Trust, and this allowed her to adopt an unusually broad approach to the study of children's disorders. Her expertise extends from neuropsychology into behaviour genetics, auditory processing, and hemispheric specialization. As well as studying children with developmental language disorder, her interests extend to encompass related conditions such as autism and dyslexia.

Dorothy has authored two books and edited four others, and published over 300 papers in scientific journals. In her prize-winning book, "Uncommon Understanding", she achieved a synthesis of work on children's comprehension disorders, relating typical and atypical development and covering processing of language right through from auditory analysis of the speech signal up to interpretation of a speaker's meaning. Her research has led to practical applications, in the form of widely-used assessments of children's language, the Test for Reception of Grammar, and the Children's Communication Checklist.

Dorothy is a Fellow of the Academy of Medical Sciences, a Fellow of the British Academy, and a Fellow of the Royal Society. Her clinical contributions have also been recognized with Honorary Fellowships from the Royal College of Speech and Language Therapists, the British Psychological Society, and the Royal College of Pediatrics and Child Health. She has Honorary Doctorates from the Universities of Newcastle upon Tyne, UK, Western Australia, Lund, Sweden, École Normale Supérieure, Paris, and Liège, Belgium. She is an Honorary Fellow of St John's College, Oxford.

Paul A. Thompson

Paul Thompson is Assistant Professor in Applied Statistics and the department lead for statistics and quantitative methods at the Centre for Educational Development, Appraisal and Research (CEDAR) at the University of Warwick.

Between October 2008 and July 2009, he worked as a postdoctoral researcher for the Fetal Medicine Foundation. This role looked at novel pre-natal screening algorithms for Down's Syndrome and pre-eclampsia. He then moved to the University of Plymouth as a Senior Research Fellow in Applied Statistics at the Centre for Biostatistics, Bioinformatics and Biomarkers, in the Schools of Medicine and Dentistry (Formerly Peninsula Medical School). This role was fully funded by a National Institute for Health Research (NIHR) Grant entitled "Clinical Trials Methods in Neurodegenerative Diseases". The focus of this research was the development of statistical models that showed individual differences in patients' disease trajectories.

Between January 2014 and September 2021, he worked as a statistician at the University of Oxford, first, with Professor Maggie Snowling (Centre for Reading and Language, University of Oxford) and then joining Professor Dorothy Bishop's research group, Oxford Study of Children's Communication Impairments, working on a range of research topics including genetics, neuroimaging, behavioural studies, methods research, and open science practices.

List of Figures

List of Tables

1

Introduction

1.1 Learning objectives

By the end of this chapter, you will be able to:

- Understand and explain the importance of evaluating interventions;
- Recognize and distinguish between examples of random error and systemic bias.

1.2 How can we know if we've made a difference?

Anthony was a 60-year-old builder who suffered a stroke that left him paralysed on his right side and globally aphasic, with difficulties producing and understanding language. He was discharged home after three weeks in hospital, by which time he recovered the ability to walk and talk, but still had severe word-finding problems. He received weekly sessions with a speech and language therapist for 6 weeks, after which his word-finding difficulties had reduced markedly. He is full of praise for the therapist and says she made a huge difference.

At two years of age, Tina's speech was markedly delayed. She had an expressive vocabulary of just ten words (mummy, daddy, doggie, water, more, want, juice, milk, bread, and bear), and her repertoire of speech sounds was limited, so what she said was often not intelligible to strangers. She was socially engaged and an assessment showed that she had age-appropriate understanding of what others said to her. The therapist worked with Tina's mother to encourage her to talk about the toys Tina was playing with and to repeat and expand on her utterances. The mother said this transformed her interactions with her daughter. Six months later, Tina's speech was much clearer and she was talking in 2-3 word utterances.

Martha is a teaching assistant based in a primary school in an area of high social deprivation. She has worked with the school's speech and language therapist to develop a language intervention programme with a class of 5-year-olds that involves regular group sessions of story-book reading with an emphasis on developing the children's vocabulary. A vocabulary test that was given at the start of the school term and again 3 months later shows that on average children know ten more of the vocabulary items after the intervention than they did at the outset. The class teacher was enthusiastic about the intervention and wants to roll it out to more classes.

These three vignettes illustrate the kinds of everyday problems confronting speech and language therapists going about their daily work: language problems are identified, interventions implemented, and, in many cases, improvements are observed. But the thoughtful therapist will have a niggling doubt: yes, in each case we see an improvement in language skills, but would this have occurred anyway? People with aphasia often recover over time, late-talkers turn out to be "late-bloomers", and children's vocabulary grows as they get older.

Readers might wonder whether we should worry. Therapists often have to work in situations of uncertainty – the clients are there and can't wait for professionals to do trials before we adopt an intervention. And in each of these cases, the speech and language therapist used their professional judgment to intervene, and an improvement in language was seen. So does it matter whether that improvement would have occurred anyway?

In our view it matters enormously, for four reasons.

- First, we owe it to those who receive our interventions to apply due diligence to ensure that, as far as is possible, what we are doing is evidence-based and unlikely to do harm – which in the best case may involve wasting people's time or money, and in the worst case could cause emotional damage. The purity of the motives of a practitioner is not sufficient to ensure this – they have to have the skills and willingness to consider the evidence dispassionately.

- For any caring professional to be taken seriously, they need to show that they evaluate the interventions they use: otherwise they may be seen as no better than purveyors of popular alternative health cures, such as aromatherapy or iridology.

- Third, someone – either the client or the taxpayer – is paying for the therapist's time. If some interventions are ineffective, then the money could be better spent elsewhere.

- Fourth, if professionals rely purely on traditional practice to determine which interventions to use, then there is no pressure to develop new interventions that may be more effective.

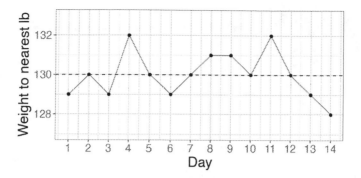

FIGURE 1.1
Random fluctuations in Bridget Jones' weight, measured to the nearest pound.

Showing that an intervention works in an individual case is very hard – especially when dealing with a condition that fluctuates or is self-limiting. In later chapters we shall consider how we can make use of group studies to evaluate interventions, and how single case designs can sometimes give greater confidence that a real effect has been achieved. But first, it is important to recognize the range of factors that conspire to make it difficult to answer the question, "Did I make a difference?" To do this, we need to understand about two factors that can mask or mimic real intervention effects: random error and systematic bias. **Random error** is used to describe influences that do not alter the average outcome, but which just make it liable to vary capriciously from one occasion to the next. **Systematic bias** refers to influences that lead outcomes to move in one direction, which may mimic an intervention effect.

1.3 The randomness of everything

Anyone who has tried to diet will be aware of how weight can fluctuate, a fact used to comic effect in the novel *Bridget Jones's Diary* (Fielding, 1996), where the protagonist's mood soared and plummeted depending on whether she was a pound heavier or lighter than the previous day. Some of the factors affecting what the scales say may be systematic and depend on calories ingested and worked off, but some will be fairly random: how you stand on the scales, whether you had a cup of tea prior to weighing, and whether the floor is level may affect the reading. And a different set of scales might give a different result. Figure 1.1 shows a notional weight chart for Bridget, who is obsessively measuring herself daily on a set of scales that weighs to the nearest pound. This plot was generated by simulating numbers with random variation, to reflect the day to day fluctuations that characterize such measurements. This is variously referred to as noise or random error in the intervention literature. Bridget may be euphoric by Day 14, as she sees a decline in her weight, but this trend is meaningless, and entirely down to the play of chance.

In effect, when we assess an intervention we aim to sift out any systematic effect of treatment from the background noise. This will be much harder if we have a measure that hops around a lot from one measurement occasion to the next regardless of our intervention – a "noisy" measure. So when we evaluate an intervention, we need to identify measures that are well-suited to assess the outcome, i.e., within minimal unwanted variation. We will discuss this further in Chapter 3.

More on random error

Note that for the purposes of an intervention study, we treat as "noise" or "random error" anything that creates variability in our measurements that is not part of the treatment effect. Strictly speaking, we can distinguish "measurement error", which relates to noise in the measurement system we are using (e.g., the scales) and other sources of noise, such as genuine fluctuations in weight from day to day that do not trend up or down over time. If Bridget hopped off and on the scales repeatedly in a 5 minute interval, there would be some fluctuations in her weight due to the scales (measurement error), especially if very precise measurement was used. But we could be confident her weight had not genuinely altered within that time period. Such variation is likely to be smaller than the variation observed day by day, which includes genuine variation around Bridget's average weight, reflecting physiological factors and gut contents. It is usually hard to isolate the causes of noisy measures: for our purposes, when we refer to noisy measures, the focus is just on how much a score varies in a random way – i.e., not consistently up or down.

One consequence of noisy measures is that they can show big changes from one occasion to another which, in individuals, might give a misleading impression of genuine treatment effects.

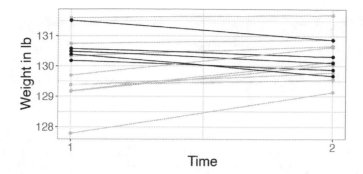

FIGURE 1.2
Baseline and outcome measures for an ineffective diet.

Figure 1.2 shows data from 14 people given a dietary intervention which does not have any beneficial effect. The data were simulated using a random number generator that just allocated each person a starting weight at random (time 1), and then created an outcome weight by adding a random number to half the starting weight (time 2). In the plot, time 1 is the 'baseline' measure, prior to the diet starting, and time 2 is the 'outcome' measure, after a period of dieting. We might be tempted to conclude that there is a subset of people who respond to the treatment – those shown in black in the figure. And indeed, those who have lost weight may be inclined to tell their friends about the marvelously effective diet, whereas those who did not remain silent. If we were to just ignore the remainder of the group, we could give a quite false impression of effectiveness of intervention. Unfortunately, this sometimes happens, when people fail to recognize that some gains in a measure could just be due to chance, and conclude instead that there is a subset of "responders" and another subset of "nonresponders" – sometimes called "treatment resisters". In fact, there will always be some variation in outcomes, and not all change is meaningful. It is possible to design a study to avoid making this basic error, but you need to be aware of it to defend against it.

1.4 Systematic bias

An even greater danger for the unwary researcher is a failure to take into account systematic bias. Systematic bias can take many forms, but of particular concern is the case where there are changes that would occur regardless of any intervention. The three examples that we gave at the outset are all cases where spontaneous change is likely even without intervention: with many kinds of brain injury, including stroke, there is very rapid recovery immediately after the injury – thought to be related to basic physiological changes such as reduction of swelling – followed by a longer period, that can persist for months or years, during which recovery continues, albeit at a slower pace. The amount and duration of such recovery can vary markedly depending on factors such as the age of the person and the nature, location, and extent of the injury.

The second example, of Tina, illustrates another case – late talkers – where substantial spontaneous improvement can occur. Despite a growing number of prospective studies of late talkers, we are still rather bad at predicting which children will be "late bloomers", who turn out fine without any special help, and which children will go on to have more persistent, long-term problems.

Finally, in the third example, we would expect children to gain new vocabulary as they get older: the difficulty is knowing just how much change it is reasonable to expect over a 3-month period as a consequence of growing older.

In all three cases, therefore, we are not only trying to detect a signal from among noise – i.e., an effect of intervention using measures that inevitably contain random error: we also want to know whether any signal we see is due to the intervention that was provided, or whether it is just part of natural change. We need to control for any form of systematic bias that could lead baseline and outcome scores to differ, regardless of the intervention. In subsequent chapters we will describe methods that have been developed that provide a rational approach to addressing this question.

It isn't always as difficult as in our three examples. Some conditions are more stable and pose less of an interpretive problem as natural recovery is unlikely. But, in our experience, research on intervention often fails to take into account systematic bias resulting from spontaneous improvement, and so we will be focusing mostly on that. And methods that are designed to deal with this kind of systematic bias can also be applied to more stable situations.

1.5   Check your understanding

In the following examples, is there random error, systematic bias, or both? (For comments on these exercises see Section 24.11).

1. We are measuring people's weight, but the scales are incorrectly calibrated, so the true measurement is always underestimated. (**systematic | random**)

2. To get an index of social responsiveness, children's interactions with others are measured in the classroom. The researcher makes one

observation per child, with different children assessed at different times of day during different activities. **(systematic | random)**

3. In an online experiment where we measure children's keypress responses on a comprehension test, the internet connection is poor and so drops out intermittently. **(systematic | random)**

4. In an intervention study with aphasic individuals, prior to treatment, vocabulary is measured on a naming test by a therapist in the clinic, and at follow-up it is measured at home by a caregiver who gives clues to the correct answer. **(systematic | random)**

2

Why observational studies can be misleading

2.1 Learning objectives

By the end of this chapter, you will be able to:

- Describe potential confounds in observational studies;

- Reflect on how these can be controlled for in experimental research designs.

Janice, a speech-and-language therapist, uses a parent-based approach with late-talking 2-year-olds. Parents make video recordings of interactions with their child, which are then analyzed with the therapist, who notes ways of making the interaction more contingent on the child's interests and communicative attempts. Janice wants to evaluate what she is doing. A colleague of hers, Anna, who has a similar caseload, is sceptical about whether Janice's approach is cost-effective. Anna uses a watchful waiting approach with children this young. Janice and Anna agree to do a study using the same pre- and post-intervention assessments so that they can evaluate the impact of Janice's approach.

Stephen notes that some aphasic patients referred to his clinic talk glowingly about a commercially available "brain-based learning" programme, MyLang-Booster. He wants to know whether he should be recommending this programme to other clients, so he carries out a comparison of patients who used MyLang-Booster and those who did not.

Dorothy was studying outcomes of 4-year-olds who had poor language skills, some of whom had received intervention. She noted that their outcomes at age 5.5 years were very variable. Some children had caught up with their peer group whereas others had persistent problems. When she presented the results, a person in the audience suggested that amount of intervention might be a factor influencing outcomes.

2.2 Confounds in observational studies

All of these vignettes illustrate **observational studies**: ones where we use naturalistic data rather than exerting any experimental control over who gets an intervention. The last one, as you may have guessed, is a true story – experience with a longitudinal study by Bishop and Edmundson (1987). Results on the relationship between intervention and outcome in this sample were disquieting to speech and language therapists: the children who had the most intervention had the worst outcomes. Did that mean that intervention was actually harming children? Well, it's possible – as we will see in Chapter 6, it is dangerous to assume that all effects of intervention are benign. But there is a much simpler explanation for this topsy-turvy result: the children who received intervention were different from those who didn't – in general they had more severe problems to start with. This makes sense: if a child is referred to speech and language therapy services, then the therapist makes a judgement about which ones to intervene with, and these are likely to be different from those who are discharged or simply reviewed after a few months. If, as appears to have been the case in the Bishop and Edmundson study in the 1980s, therapists work most with the more serious cases, then there is a **confound** between severity of language problems and receipt of therapy, and this can make intervention appear ineffective. This gives a very clear illustration of the perils of observational studies.

Bias can also work in the opposite direction: a contemporary concern in the UK is that therapists are being encouraged to focus their efforts on children who can be discharged after a short block of treatment, which means they tend to work with children with milder problems (Ebbels et al., 2019). This will create the opposite impression – therapy which is actually ineffective may appear effective. The basic problem is the same: treated and untreated groups are not comparable, and so comparing them will give misleading results.

Stephen's study is equally problematic. Here we are comparing a self-selected group of patients with his regular caseload. Those who tried MyLangBooster may be more motivated to improve than other patients. They may have more money, so they can afford to pay for the programme. Or they may be more desperate – having tried other interventions that failed to make a difference. Furthermore, Stephen may only hear from those who felt they improved and be unaware of other patients who tried it but then dropped out because they obtained disappointing results. It is almost never a good idea to base an evaluation of an intervention on a study of self-selected enthusiasts. There are just too many potential confounds that could cause bias.

What about the case of Janice and Anna? This may seem less problematic, since the two therapists have similar caseloads, and the decision about therapy

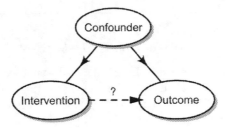

FIGURE 2.1
Path diagram showing confounder influence.

is based on therapist preference rather than child characteristics. Here again, though, the comparison has the potential to mislead. If baseline and outcome assessments are done using the same measures, then it is at least possible to check if the children in the two practices are similar at the outset. But there would still be concerns about possible differences between the therapists and their practices that might be influencing results. Maybe Anna rejects parent-based intervention because she knows that most of the parents in her practice have full-time jobs and would not be willing or able to devote time to attending sessions to analyze videos. Maybe Janice is an exceptionally skilled therapist who would obtain good outcomes with children regardless of what she did. Perhaps her enthusiasm for a parent-based approach contrasts with Anna's more downbeat attitude, and this has an influence on parent and/or child. In sum, there is scope for non-specific treatment effects to exert an impact. If Janice finds better outcomes than Anna, despite both therapists doing their best to ensure that the participating children and parents from their practices are similar, then it is reasonable to say that this is useful information that would provide justification for a more formal comparison. But it is not conclusive and cannot substitute for the kind of experimental study discussed in the next chapter.

Things that may affect outcomes and that differ between intervention and control groups are known as **confounders**. The relationship between confounders and other variables can be depicted in the kind of **path diagram** shown in Figure 2.1. A path diagram is a formal depiction of the pattern of relationship among variables, where an arrow indicates the direction of a causal influence. In this diagram, the confounder on the one hand influences who gets intervention and on the other hand affects the outcome. Here we have shown the direct relationship from intervention to outcome by a dashed arrow with a question mark, to indicate that it is uncertain whether there is a genuine benefit from intervention. Even if there is no such influence, the confounder can give the impression that the intervention is effective, because it induces a positive association between intervention and outcome.

Consider possible confounders in the following examples: Does long-term use of hormone replacement therapy (HRT) carry risks or benefits? Does excessive use of computer games in teenagers cause social isolation? Will your life be extended if you eat more cabbage? Here are just a few possibilities: Women who decide to continue to use HRT may have more severe menopausal symptoms than those who do not. Excessive use of computer games may be a consequence rather than a cause of lack of social engagement, and those who eat cabbage may adopt healthier lifestyles than those who don't.

Many reports in the media about things that are good or bad for you are based on observational rather than experimental data. In some cases, it is hard to see how other types of data could be collected: are we really going to succeed in allocating adolescents to a regime of non-stop computer games, or to force people to start eating cabbage? Indeed, some associations that are now well-established, such as the link between cigarette smoking and cancer, were discovered from observational data, and could not be studied any other way. But where the interest is in interventions administered by a therapist, then it should not be necessary to rely on observational studies, and, as we have shown, to do so can lead to flawed conclusions

There is a great deal of work on methods for analyzing and interpreting observational data (see, e.g., Hernan (2018)), but, given the difficulty of identifying causal influences from such evidence, we will not say any more about this topic, but rather focus on experimental approaches that allow us to minimize the role of confounders.

2.3 Check your understanding

1. Find a newspaper report of a factor that is reported to be a risk or benefit to health. Is it based on an experimental or observational study? Can you identify potential sources of bias?

2. Figure 2.2 is recreated from a medical example by Rothwell (2005), and shows the rate of stroke or death at 5 year follow-up in at-risk patients who either underwent a surgical procedure known as endarterectomy, or who had medical treatment as usual.

The two bars on the left show results obtained in an experimental study where assignment to surgery was done at random, whereas the bars on the right show results from an observational study of patients who were not included in a trial. What do these plots show, and how might it be explained?

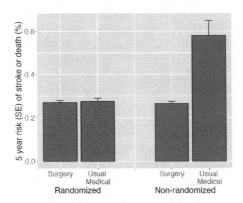

FIGURE 2.2

Outcome from surgery in randomized vs. non-randomized studies.

3

How to select an outcome measure

3.1 Learning objectives

By the end of this chapter, you will be able to:

- Understand the key concepts of reliability, validity, sensitivity, and efficiency;
- Evaluate test content in relation to research goals.

3.2 Criteria for selecting a good measure

Suppose you want to evaluate the effectiveness of a parent-based intervention for improving communication in three-year-olds with poor language skills. You plan to assess their skills before the intervention, immediately after the intervention, and again six months later. The initial measurement period is known as the **baseline** – because it acts as a reference point against which improvement can be measured.

There are many measures you could choose: the child's mean length of utterance (MLU) from a language sample, scores on a direct assessment such as preschool CELF (Wiig et al., 2006), the parent's response to language items on a screening inventory such as the Ages and Stages Questionnaire (Bricker & Squires, 1999). You may wonder whether you should include as many measures as possible to ensure you cover all bases. However, as we shall see in Chapter 14, if you measure too many things, you will need to make statistical adjustments to compensate for the risk of getting spurious positive results, so it is often recommended to specify a primary outcome measure – the one you would put your money on as most likely to show the effect of interest, if you were a betting person.

The key questions to ask are:

1. Is the measure reliable?

2. Is it valid? i.e., does it measure what I want to measure?

3. Are there norms?

4. Is it sensitive?

5. Is it efficient?

Types of numerical measurement

Results from a measurement in an intervention study are typically represented as numbers, but different types of measurement have different properties. **Nominal**, also known as categorical, data are just used to designate qualitative information. For instance, we may record data as 1, 2, or 3 to indicate the school that a child attends. The order of the numbers here is arbitrary, and the information could equally well be represented by letters, e.g., school A, B, and C. It would make no sense to perform numerical operations such as addition and subtraction on such numbers. For **ordinal** data, the order of the numbers matters. For instance, if we had a 3 point scale of fluency, a person with a score of 3 would be more fluent than one with a score of 2, who would in turn be more fluent than a person who scored 1. In effect ordinal data describe ordered categories, but again, you can't sensibly perform arithmetical operations on such data, though you can specify a median or mode to represent the most typical score. Data on an **interval scale** are commonly found in behavioural studies. Here, the numbers are equally spaced, so the difference between 1 and 2 is the same quantity as the difference between 3 and 4. With interval data, we can compute an average and conduct many types of statistical tests that are not possible with ordinal data. The most sophisticated type of numerical measurement is the **ratio scale**, which has all the properties of interval data, plus it has a meaningful zero point. Weight, height, length, and duration are all measured on a ratio scale: a zero measurement indicates there is no weight, height, length, or duration. On this kind of scale, one can meaningfully interpret multiplication – e.g., 4 kg is twice as much as 2 kg. Most commonly used statistical procedures assume data are at least on an interval scale. If you are using ordinal data, then a different kind of test using *nonparametric* statistics is required.

3.3   Reliability

You may be surprised to see reliability at the top of the list. Surely validity is more important? Well, yes and no. As shown in Figure 3.1, there's not much

FIGURE 3.1
Reliability and validity illustrated as a sharpshooter hitting a target.

point in having a measure that is valid unless it is also reliable: what we want is a measure that hits the bullseye, in terms of measuring what we are aiming to measure, not just on average, but consistently. A measure that is valid but not reliable is worse than useless in an intervention study, so we put reliability at the top of the list.

So what is reliability? This brings us back to the issue of random error or "noise": quite simply you want a measure that is as accurate as possible, with minimal influence from random, unwanted sources of variation. One way to assess this is to apply the same measure on two occasions (without any intervention between) to see how similar they are, computing an index of "test-retest reliability". To assess this, we repeat the measurement with a sample of individuals seen on two occasions close in time (i.e., before we expect any change due to maturation or intervention). Test-retest reliability is often quantified by the correlation coefficient (r). The correlation coefficient is a statistic that takes the value of zero when there is no relationship between two variables, 1 when there is a perfect relationship, and -1 when there is an inverse relationship. Figure 3.2 shows scatterplots illustrating the relationship between scores at Time 1 and Time 2 in the case of perfect correlation ($r - 1$), high correlation ($r = .8$), low correlation ($r = .25$), and no correlation ($r = 0$). If you draw a straight line through the points on the graph, the strength of correlation is reflected in how closely the points cluster around the line. Where the correlation is 1, you can predict perfectly a person's score on Y if you know their score on X.

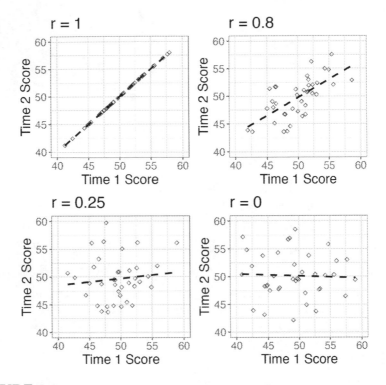

FIGURE 3.2
Scatterplots showing relationship between scores at Time 1 and Time 2 for 40
individuals (simulated data).

Correlation as an index of reliability

Although the correlation coefficient is widely used to index reliability,
it can be misleading because it just reflects agreement in the rank
ordering of two measures. It would be possible to get a high correlation even if the task showed practice effects, so that scores at Time 2
were substantially higher than those at Time 1. For that reason, the
intraclass correlation coefficient (ICC) is preferable, as this reflects
agreement in absolute scores. In practice, however, the two types of
correlation give similar values unless there is a substantial difference
in means between the two occasions of measurement. For more explanation see https://www.measurement-toolkit.org/concepts/statistical-
assessment (NIHR Cambridge Biomedical Research Centre, n.d.).

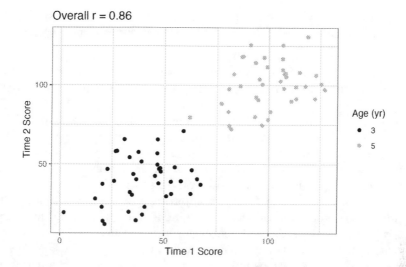

FIGURE 3.3
Scatterplot showing relationship between Time 1 and Time 2 scores for children from two age bands (simulated data).

So how reliable should a measure be? Most psychometric tests report reliability estimates, and a good test is expected to have test-retest reliability of at least .8. But be careful in interpreting such estimates, as you need also to consider the age range on which the estimate is based.

Figure 3.3 shows how a test of vocabulary size that looks highly reliable when considered in a combined sample of 3-year-olds and 5-year-olds is really not very reliable when we just look at a single year-band. Although the overall correlation for Time 1 and Time 2 is .86, within each age band it is only .3. This is because the index of reliability, the correlation coefficient, is affected by the range of scores: older children tend to score higher than younger children, and this is reflected in the correlation. If your study was focused just on 3-year-olds, you'd really want it to be reliable *within* that age range.

The topic of reliability is covered more formally in classical test theory (Lord et al., 1968) and subsequent developments from this. These involve a mathematical approach that treats an observed test score as the sum of a "true" effect (i.e., what you want to measure) plus random error.

Most widely-used language assessments have been developed using methods from classical test theory (Denman et al., 2017). However, mathematical treatments of test development have come a long way in the past 20 years, and an approach known as Item Response Theory is now generally preferred for developing more precise and valid instruments (Reise et al., 2005). This is a complex topic that is beyond the scope of this book. The most important message to take away is that the lower the reliability, the greater the random error, and the harder it is to detect the true effect of intervention against the background of noise.

We have focused on test-retest reliability as this is the most relevant form of reliability in intervention studies. If you plan to use the same measure at baseline and as an outcome measure, then what you need to know is how much variation in that measure you might expect if there were no effect of intervention. There are other reliability indices that are sometimes reported with psychometric tests. Split-half reliability and internal consistency (Cronbach's alpha) both consider the extent to which a score varies depending on the specific items used to calculate it. For instance, we could assess split-half reliability for mean length of utterance (MLU) by computing scores separately for all the odd-numbered utterances and the even-numbered utterances and measuring the correlation between the odd and even scores. Although this gives useful information, it is likely to differ from test-retest reliability, because it does not take into account fluctuations in measurement that relate to changes in the context or the child's state.

It is much easier to compute measures of internal consistency than to do the extra testing that is needed to estimate test-retest reliability, and some published psychometric tests only provide that information.

A recent review by Denman et al. (2017) looked at psychometric properties, including reliability, of 15 comprehensive language assessments for children aged 4-17 years. Although all but one reported test-retest reliability that was rated as good, the methodological quality of the psychometric data was deemed questionable for all but two tests.

3.4   Validity

A modification of a popular adage is "If a thing is not worth doing, it's not worth doing well." This applies to selection of outcome measures: you could have a highly reliable measure, but if it is not measuring the right thing, then there's no point in using it.

Deciding what is the "right thing" is an important part of designing any invention study, and it can be harder than it appears at first sight. The answer might be very different for different kinds of intervention. One consideration of major importance when dealing with speech and language assessments is the cultural context.

3.4.1   Cultural factors

The first author still remembers her first encounter with a popular receptive vocabulary test, the Peabody Picture Vocabulary Test, where the task was to match a spoken word to one of four pictures in an array. An early item was "caboose", a word she had never encountered, but which was presumably familiar to children in the USA. Another American reading assessment required children to find words that sounded the same, and treated 'aunt' and 'ant' as homophones – words with completely different vowels in standard British English. Conversely, in standard British English, the words "arm" and "calm" have the same ending, whereas to speakers of Scots, Irish, and American English, the 'r' is pronounced. In some dialects, low intensity morphemes may be omitted, and it's acceptable to say "She my sister" rather than the standard English "She's my sister", or to say "I done my homework" rather than "I've done my homework". And if we move to consider pragmatics and language usage, there can be wide variability from culture to culture in what is deemed acceptable or odd in terms of what is said and how it is said.

It is crucial to be sensitive to such issues. Quite simply, an assessment that categorizes nonstandard language as impairment is invalid. Terms such as language impairment or disorder are only appropriate when the individual is recognized as having problems within their own cultural community. We therefore need to use culturally appropriate assessments that do not confuse language difficulties with normal variation. While that is obvious advice to give, there may be a lack of suitable assessments available, in which case some modification of existing assessments may be the only option. This should be done in consultation with someone who is a member of the relevant culture. If modifications are made, then normative data (see below) should be interpreted with extreme caution.

3.4.2   Practice effects

If you repeatedly do the same task, you may improve just because of increased familiarity with the task. We defer discussion of practice effects to Chapter 5, but note here that we usually will want to use outcome measures that show minimal benefits of practice. This is particularly the case when using single-case designs (see Chapter 20), where the same assessment may be repeatedly administered to the individual on several occasions.

3.4.3   Generalizability of results: the concepts of far and near transfer

Generalizability is an issue that is particularly relevant to the first and third vignettes from Chapter 1, word-finding intervention for aphasia and the classroom-based vocabulary intervention. These vignettes illustrate interventions that have a specific focus. This means we can potentially tie our outcome measures very closely to the intervention: we would want to measure speed of word-finding in the first case and vocabulary size in the second.

There is a risk, though, that this approach would lead to trivial findings. If we did a word-finding training with an aphasic client using ten common nouns and then showed that his naming had speeded up on those same ten words, this might give us some confidence that the training approach worked (though we would need appropriate controls, as discussed in later chapters). However, ideally, we would want the intervention to produce effects that generalized and improved his naming across the board. Similarly, showing that a teaching assistant can train children to learn ten new animal names is not negligible, but it doesn't tell us whether this approach has any broader benefits.

These issues can be important in situations such as phonological interventions, where there may be a focus on training the child to produce specific contrasts between speech sounds. If we show that they master those contrasts but not others, this may give us confidence that it was the training that had the effect, rather than spontaneous maturation (see Chapter 4), but at the same time we might hope that training one contrast would have an impact on the child's phonological system and lead to improved production of other contrasts that were not directly trained.

These examples illustrate the importance of testing the impact of intervention not only on particular training targets but also on other related items that were not trained. This is something of a two-edged sword. We may hope that treatment effects will generalize, but if they do, it can be difficult to be certain that it was our intervention that brought about the change. The important thing when planning an intervention is to think about these issues and consider whether the mechanism targeted by the treatment is expected to produce generalized effects, and if so to test for those.

The notion of generalization assumes particular importance when the intervention does not directly target skills that are of direct relevance to everyday life. An example is Cogmed Working Memory Training, a computer-based intervention that has been promoted as a way of improving children's working memory and intelligence. The child plays games that involve visual tasks that tax working memory, with difficulty increasing as performance improves. Early reports maintained that training on these tasks led to improvement on nonverbal intelligence. However, more recent literature has challenged this claim, arguing that what is seen is "near transfer" – i.e., improvement in the types of memory task that are trained – without any evidence of "far transfer" – i.e., improvement in other cognitive tasks (Aksayli et al., 2019). This is still a matter of hot debate, but it seems that many forms of "computerized brain training" that are available commercially give disappointing results. If repeatedly doing computerized memory exercises only improves the ability to do those exercises, with no "knock on" effects on everyday functioning, then the value of the intervention is questionable. It would seem preferable to train skills that would be useful in everyday life.

3.4.4   Functional outcomes vs. test scores

The second vignette (Tina) in Chapter 1 focuses on an intervention where issues of far and near transfer are not so relevant, as the intervention does not target specific aspects of language but rather aims to modify the parental communicative style in order to provide a general boost to the child's language learning and functional communication. This suggests we need a rather general measure; we may consider using a standardized language test because this has the advantage of providing a reasonably objective and reliable approach to measurement. But does it measure the things that clients care about? Would we regard our intervention as a failure if the child made little progress on the standardized test, but was much more communicative and responsive to others? Or even if the intervention led to a more harmonious relationship between parent and child, but did not affect the child's language skills?

We might decide that these are important key outcomes, but then we have to establish how to measure them. In thinking about measures, it is important to be realistic about what one is hoping to achieve. If, for instance, the therapist is working with a client who has a chronic long-term problem, then the goal may be to help them use the communication skills they have to maximum effect rather than to learn new language. The outcome measure in this case should be tailored to assess this functional outcome, rather than a gain on a measure of a specific language skill.

If you are devising an intervention study, it can be useful to consult with representatives of the clinical group you plan to work with, to see what they regard as the most useful outcomes. This kind of stakeholder engagement can be valuable in increasing the impact of intervention research – see Forsythe et al. (2019) for examples.

3.4.5   Subjectivity as a threat to validity

In later chapters we will discuss various sources of bias that can affect studies, but one that crops up at the measurement stage is the impact of so-called "demand characteristics" on subjective ratings. Consider, for a moment, how you respond when a waiter comes round to ask whether everything was okay with your meal. There are probably cultural differences in this, but the classic British response is to smile and say it is fine even if it was disappointing. We tend to adopt a kind of "grade inflation" to many aspects of life when asked to rate them, especially if we know the person whose work we are rating.

In the context of intervention, people usually want to believe that interventions are effective and they don't want to appear critical of those administering the intervention, and so ratings of language are likely to improve from baseline to outcome, even if no real change has occurred. This phenomenon has been investigated particularly in situations where people are evaluating treatments that have cost them time and money (because of a phenomenon known as cognitive dissonance), but it is likely to apply even in experimental settings when interventions are being evaluated at no financial cost to those participating.

In the field of pharmaceutical trials, there is a simple solution, which is to keep the patient unaware of whether or not they are receiving an active drug or a placebo. This kind of masking is what is referred to as a "single blind trial". However, in most contexts where a behavioural intervention is used, it is not possible to keep people unaware of what intervention condition they are in. This means we have to be particularly vigilant not to rely too heavily on subjective assessments by those who receive the intervention.

An example in the published literature comes from Loeb et al. (2001) who did a small-scale study to evaluate a computerized language intervention, FastForword (FFW). This study is noteworthy because as well as measuring children's language pre- and post- intervention, it included parent ratings of children's outcomes. There was, however, a striking dissociation between the reports of parental satisfaction with the intervention and the lack of improvement on language tests. Another example comes from a trial by Bull (2007) of "Sunflower therapy" for children with a range of neurodevelopmental disorders; here again parents were very positive about the intervention, but objective measures showed children had made no significant progress relative to a control group.

Such results are inherently ambiguous. It could be that parents are picking up on positive aspects of intervention that are not captured by the language tests. For instance, in the Sunflower therapy study, parents reported that their children had gained in confidence – something that was not assessed by other means. However, there it is hard to know whether these evaluations are valid, as they are likely to be contaminated by demand characteristics.

Ideally we want measures that are valid indicators of things that are important for functional communication, yet are reasonably objective – and they need also to be reliable and sensitive! We don't have simple answers as to how this can be achieved, but it is important for researchers to discuss these issues when designing studies to ensure they use optimal measures.

3.4.6   Correlations with other measures

The usual approach to assessing validity involves demonstrating that a new test instrument gives results that correlate with other existing measures of the same construct. Usually one sees moderate correlations between old and new measures, which may offer some reassurance that they are in part measuring the same thing while each contributing more specific information.

We have emphasized other aspects of validity, as our view is that validity has to be considered in terms of how valid a measure is for the purpose to which it will be put, rather than by some conventional yardstick. It can be helpful to know how far two tests give similar results, but this information is not all that informative unless we have a true "gold standard" against which to compare a measure. If the measures are very highly correlated, then one might as well stick with the original measure rather than developing a new version, unless our new version is more efficient (e.g., takes less time to administer or score). If they are not at all correlated, it could mean that one of the measures is a poor indicator of the underlying construct, but we do not know which one.

Where two instruments claim to measure the same construct, yet are only moderately correlated, this can provide insight into factors that influence a score. Frizelle et al. (2019) compared scores on two assessments of comprehension of complex sentences. In one task, the child was asked to match a spoken sentence to an item from an array of four pictures; in another task, the child was shown an animation and was asked to judge whether or not it was described by a spoken sentence. The latter task was easier for younger children, and the order of difficulty of sentence types was different in the two tasks. One reason was because the sentence-picture matching task requires children to scan an array of pictures while keeping a spoken sentence in mind. Children who are inattentive, impulsive, or have poor short-term memory may do poorly on this task, even if their language understanding is reasonable. The researchers concluded: "We suggest that those designing and administering language assessments need to critically appraise the tools they are using and reflect on whether the test is actually measuring what it claims to." There is probably no perfect assessment: there will always be unwanted factors that influence our measures. But if we understand what these are, it may be possible to take them into account.

3.5   Normative data

A common approach in test development is to gather data on a standardization sample, i.e., a group of individuals who will provide a yardstick against which a test score can be evaluated.

A standardization sample provides normative data that are useful for two reasons. First, they indicate that the test has been applied to a large sample of individuals, and that gives confidence that basic problems with the test have been ironed out – e.g., items that are ambiguous will be detected and can be removed. Second, if there is an adequate normative sample, it is possible to convert raw scores (e.g., the number of items correct) to scaled scores, which give an indication in statistical terms of how a person's performance on the assessment compares with the standardization sample. This can be particularly

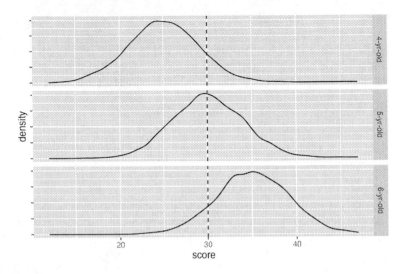

FIGURE 3.4

Density plot showing distribution of vocabulary scores for three age groups (simulated data). The dotted line shows a raw score of 30, which is above average for 4-year-olds, average for 5-year-olds, and below average for 6-year-olds.

useful when we are using assessments that are age-dependent. We explain here how scaled scores are derived, but as we will argue later, in the context of an intervention study, it is not always necessary or desirable to use scaled scores.

Let us suppose we have gathered vocabulary test data on 100 children for each year band from 4 to 6 years. Figure 3.4 shows the distributions of scores at each year band – these data are normally distributed in a bell curve, which means each age band can be summarized by two statistics – the mean (or average) and the standard deviation (SD), which is a measure of the spread of the data. For these simulated data, the means for the three groups are 25, 30, and 35, and each group has a SD of 4. The z-score indicates how many standard deviations a raw score is above or below average. We can convert a raw score (e.g., a score of 30 correct) into a z-score by subtracting the mean and dividing by the SD. For the data shown in Figure 3.4, a raw score of 30 is above average for a 4-year-old (z-score of 1.25), average for a 5-year-old (z-score of 0), and below average for a 6-year-old (z-score of -1.25). The z-score is a kind of age-scaled score, but psychometric tests often convert the z-score to a different scale for convenience.

> ## Scaled scores and percentiles
>
> The most basic way of representing a scaled score is as a z-score. As explained in the text, this just represents distance from the mean in standard deviation (SD) units. Z-scores, however, involve both positive and negative numbers and fractional units. When entering data or writing reports, it is much easier to deal with whole numbers. So most psychometric tests simply transform the z-score onto a scale with a more convenient mean and SD. The best-known example is the IQ test, which usually transforms the scaled score to a mean of 100 and standard deviation of 15, known as a standard score. So if someone scores one SD below the mean, this is represented as a standard score of 85; if two SD below the mean, then the standard score is 70. As if this was not complicated enough, a scaled score can also be directly translated into a percentile (or centile) score, which corresponds to the percentage of people in the relevant population who are expected to obtain a score at least as high as this. For normally distributed data, there is an orderly relationship between percentiles and the z-score: approximately 16% of people are expected to score 1 SD below the mean, and 3% score 2 SD below the mean. You may wonder why bother with scaled scores, given that percentiles have a more intuitive interpretation. Percentiles are good for communicating how exceptional a given score is, but they are not so well-suited for doing statistical analyses, as they are not normally distributed.

We have shown how norms can be used to transform the same raw score into different scaled scores, depending on age. The transformation can also achieve the opposite: for the example shown in Figure 3.4, we have 4-, 5- and 6-year-old children with raw scores of 25, 30, and 35 respectively; they would all have the same age-scaled score (z-score of zero). In effect, using age-scaled scores allows us to remove some of the noise from our measures by taking out the effect of age and representing a score purely in terms of statistical difference from the mean of the relevant age band.

So what should we be looking for in terms of test norms? As with other aspects of test criteria, it will depend upon what you want to do with a measure. In the context of intervention studies, use of scaled scores may be non-optimal because there is a loss of sensitivity. The main benefit of using scaled scores is that they provide a way of indicating how exceptional a raw score is, regardless of age. So, for example, if you were aiming to move a child's vocabulary from below average to the normal range, you could tell from a z-score if you achieved that goal, whereas a raw score would not be informative.

If one is going to use scaled scores, it is important to consult the test manual to check that an appropriate standardization sample was used. As noted above, if the test norms come from individuals from a different dialect or communicative culture, then the test may include inappropriate items that make it invalid as an index of language ability. There can be more subtle considerations. To give an example, the Test of Word Reading Efficiency (Torgesen et al., 1999) is widely used as a quick measure of word reading accuracy. However, the norms are representative of children in the USA, where reading instruction does not start until 6 years of age. This makes it unsuitable for 6-year-olds in the UK, who are introduced to reading at 4 years of age.

It is seldom possible to find a test that is standardized on a population that is a perfect match for a study sample, and so some compromises must be made; the important point is that when selecting an instrument, one should check out standardization information in the test manual and be aware that a measure might give biased scaled scores if there are systematic differences between the demographics of the normative sample and a study sample. A related concern is when a test's norms are based on samples that are either very small or selected by convenience rather than intended to be representative of the population, or where scaled scores are derived by a z-score formula based on underlying distributions that are skewed (e.g., when there are ceiling effects).

The bottom line is that in clinical contexts, scaled scores are typically more informative than raw scores for characterizing how an individual compares to others in the population, and it is important for clinicians to understand how scaled scores are derived and what they mean. In intervention studies, however, this may not be the case.

3.6   Sensitivity

Those who develop psychometric tests often focus on reliability and validity but neglect sensitivity. Yet sensitivity is a vital requirement for an outcome measure in an intervention study. This refers to the grain of the measurement: whether it can detect small changes in outcome. Consider Bridget Jones on a holiday to a remote place where there are no scales, just a primitive balance measure that allows her to compare herself against weights of different sizes. She would be unable to detect the daily fluctuations in pounds and only be able to map her weight change in half-stone units. She could genuinely lose weight but be unaware of the fact.

Many standardized tests fall down on sensitivity, especially in relation to children scoring at the lower end of the ability range. It is customary for assessment purposes to convert raw scores into scaled scores on these tests. As noted above, this allows us to have a single number that can be interpreted in terms of how well the child is performing relative to others of the same age. But these often reduce a wide range of raw scores to a much smaller set of

TABLE 3.1

Conversion from raw to scaled scores for a CELF subscale in 10-year-olds

Raw	Scaled	Raw	Scaled
0-34	1	**48**	8
35-37	2	**49**	9
38-40	3	**50**	10
41-43	4	**51**	11
44-45	5	**52**	12
46	6	**53**	13
47	7	**54**	14

scaled scores, as illustrated in Table 3.1, which shows conversion from raw to scaled scores (with mean 10 and SD of 3) for a subscale from an old version of Clinical Evaluation of Fundamentals in 10-year-olds. A child whose raw score improved from 35 to 40 would only show a one-point improvement in the scaled score, from 2 to 3. On the other hand, at the top end of the scale, a 5-point improvement would translate to a 5-point gain in scaled score. This scale also has the disadvantage of showing a ceiling effect, which means it could not demonstrate improvement beyond a scaled score of 14. Scaled scores may, therefore, be blunt instruments when the focus is on measuring small changes that are associated with intervention.

Problems with sensitivity can also be an issue with measures based on rating scales. For instance, if we just categorize children on a 5-point scale as "well below average", "below average", "average", "above average", or "well above average", we are stacking the odds against showing an intervention effect – especially if our focus is on children who are in the bottom two categories to start with. Yet we also know that human raters are fallible and may not be able to make finer-grained distinctions. Some instruments may nevertheless be useful if they combine information from a set of ratings.

Although we need sensitive measures, we should not assume that a very fine-grained measure is always better than a coarser one. For instance, we may be measuring naming latency in aphasic patients as an index of improvement in word-finding. It's unlikely that we need millisecond precision in the timing, because the changes of interest are likely to be in the order of tenths of a second at most. While there's probably no harm in recording responses to the nearest millisecond, this is not likely to provide useful information.

3.7   Efficiency

Intervention research is usually costly because of the time that is needed to recruit participants, run the intervention, and do the assessments. There will always be pressures, therefore, to use assessments that are efficient and provide key information in a relatively short space of time.

Efficiency is not always adequately considered when designing an intervention study. A common experience is that the researchers want to measure everything they can think of in as much detail as possible. This is understandable: one does not want to pick the wrong measure and so miss an important impact of the intervention. But, as noted above, and discussed more in Chapter 14, there is a danger that too many measures will just lead to spurious findings. And each new measure will incur a time cost, which will ultimately translate to a financial cost, as well as potentially involving participants in additional assessment. There is, then, an ethical dimension to selection of measures: we need to optimize our selection of outcome measures to fulfill criteria of reliability, sensitivity, and validity, but also to be as detailed and complex as we need but no more.

The first author's interest in efficiency of measurement may be illustrated with a vignette. Bishop & Edmundson (1987) conducted a longitudinal study of 4-year-olds with developmental language disorders. This was not an intervention study: rather, the goal was to identify how best to predict outcomes. When experts were asked what measures to use, a common piece of advice was to take a language sample, and then analyze it using LARSP (Crystal et al., 1977), which at the time was a popular approach to grammatical analysis.

In practice, however, language sampling seemed to provide little useful information in relation to the time it took to gather and transcribe the sample. Many of the children in the study said rather little and did not attempt complex constructions. It was possible to get more information in five minutes with two elicitation tasks (Renfrew, 1967, 2010) than from 30 minutes of language sampling. Furthermore, after investing many hours of training in LARSP, analyzing the results, and attempting to extract a quantitative measure from this process, we ended up with something that had a correlation of greater than .9 with the much simpler measure of mean length of utterance (MLU). The lesson learned was that the measure needs to fit the purpose of what you are doing. In this case, we wanted an index of grammatical development that could be used to predict children's future progress. The Renfrew tasks, which were treated dismissively by many therapists, who regarded them as too old-fashioned and simple, were among the most effective measures for doing that. A practitioner working with a child might well find LARSP and language sampling preferable for identifying therapy targets and getting a full picture of the child's abilities, but for the purposes of this study, language sampling provided far more detail than was needed.

There are other cases where researchers do very complex analysis in the hope that it might give a more sensitive indicator of language, only to find that it is highly intercorrelated with a much simpler index. In the domain of expressive phonology, the first author spent many hours developing an (unpublished) index of severity based on analysis of phonological processes, only to find that this was entirely predictable from a much simpler measure of percentage consonants correct. Here again, evaluation of phonological processes may be useful for other purposes, such as planning therapy, but it seems unnecessarily complicated if one only wants a measure of severity.

A related point is that researchers are often tempted by the allure of the new, especially when this is associated with fancy technology, such as methods of brain scanning or eye-tracking. Be warned: these approaches yield masses of data that are extremely complex to analyze, and they typically are not well-validated in terms of reliability, sensitivity, or validity! Even when high-tech apparatus is not involved, the newer the measure, the less likely it is to be psychometrically established – some measures of executive functioning fall in this category, as well as most measures that are derived from experimental paradigms. Clearly, there is an important place for research that uses these new techniques to investigate the nature of language disorders, but that place is not as outcome measures in intervention studies.

On the basis of our experience, we would advise that if you are tempted to use a complex, time-consuming measure, it is worthwhile first doing a study to see how far it is predictable from a more basic measure targeting the same process. It may save a lot of researcher time, and we owe it to our research participants to do this due diligence to avoid subjecting them to unnecessarily protracted assessments.

3.8 Check your understanding

1. A simple way to measure children's language development is in terms of utterance length. Roger Brown's (Brown, 1973) classic work showed that in young children Mean Length of Utterance in morphemes (MLU) is a pretty good indicator of a child's language level; this measure counts each part of speech (morpheme) as an element, as well as each word, so the utterance "he wants juice" has 4 morphemes (he + want + s + juice). Brown's findings have stood the test of time, when much larger samples have been assessed: see

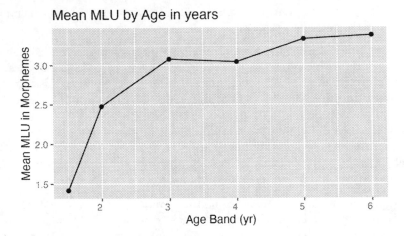

FIGURE 3.5
Mean length of utterance (MLU) values for a cohort of 630 children speaking
North American English in the Child Language Data Exchange System archive.
Recreated from Ratner and MacWhinney (2016) (data kindly provided by
authors)

Figure 3.5 from Ratner & MacWhinney (2016). Is this evidence of
reliability, validity, and/or sensitivity?

2. Here we focus on **reliability**, i.e., the likelihood that you might
   see similar results if you did the same assessment on two different
   occasions. How might you expect the reliability of MLU to depend
   on:

   - Length of language sample?

   - Whether the child is from a clinical or typically-developing
     sample?

   - Whether the language sample comes from an interaction with
     a caregiver vs. an interaction with an unfamiliar person?

3. Do a literature search to find out what is known about test-retest
   reliability of MLU. Did your answers to question 2 agree with the
   published evidence? N.B. At the end of Chapter 21, there is some
   advice on how to search the literature. If you do not have access to
   a university library you may not be able to obtain all the articles
   you find, but some will be "open access", which means anyone can
   download and read them. For those that are not open access, you
   can request a copy from the author by email: this usually elicits a
   positive response.

4. You have a three-year-old child with an MLU in morphemes of 2.0. Is there enough information in Figure 3.5 to convert this to a standard score?

5. Take one of the vignettes from Chapter 1 and consider what measures you might use to evaluate whether the intervention was effective. What are the pros and cons of different measures? How far do they meet requirements of reliability, sensitivity, validity, and efficiency?

4

Improvement due to nonspecific effects of intervention

4.1 Learning objectives

By the end of this chapter, you will be able to:

- Be aware of non-specific intervention effects and how these can be separated from the specific impact of an intervention;

- Understand the concept of mechanism measures, as distinct from broader outcome measures.

4.2 Placebo effects

Most of us are familiar with the **placebo effect** in medicine: the finding that patients can show improvement in their condition even if given an inert sugar pill.

Figure 4.1 shows outcomes of a simulated drug trial that compared three groups of people ('arms' of the trial) who were given either no treatment, an inert pill (placebo), or an active pill (intervention). We see a sizable difference between no treatment and placebo. To evaluate the true treatment effect, over and above the placebo effect, we need to compare the outcome of the intervention condition with the outcome of the placebo condition.

There is much debate about the nature of placebo effects – whether they are mostly due to the systematic changes discussed in Chapter 2 or something else. They are thought to operate in cognitive and behavioural interventions as well: communication may improve in a person with aphasia because they have the attention of a sympathetic professional, rather than because of anything that professional does. And children may grow in confidence because they are made to feel valued and special by the therapist.

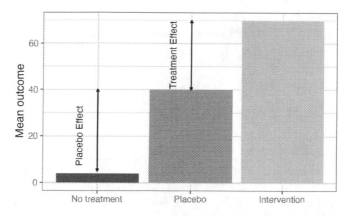

FIGURE 4.1
Placebo effect in a three-arm parallel group trial.

The Hawthorne Effect

The placebo effect in behavioural interventions is often discussed in conjunction with the Hawthorne effect, which refers to a tendency for people to change their behaviour as a result of simply being the focus of experimental attention. The name comes from the Hawthorne Works in Cicero, Illinois, where a study was done to see the effects of lighting changes and work structure changes such as working hours and break times on worker productivity. These changes appeared to improve productivity, but subsequent work suggested this was a very general effect that could be produced by almost any change to working practices. However, there is a twist in the tale. The original factory studies on the Hawthorne effect have been described as "among the most influential experiments in social science". Yet, like so many classic phenomena in social science, the work attracted critical scrutiny years after it was conducted, and the effect turns out to be far less compelling than was previously thought. The paper "Was there really a Hawthorne effect at the Hawthorne plant?" (Levitt & List, 2011) describes how the authors dug out and reanalyzed the original data from the study, concluding that "an honest appraisal of this experiment reveals that the experimental design was not strong, the manner in which the studies were carried out was lacking, and the results were mixed at best...".

Most therapists would regard such impacts on their clients as entirely valid outcomes of the intervention – boosting functional communication skills and confidence are key goals of any intervention package. But it is important, nevertheless, to know whether the particular intervention adopted has a specific effect. Many individualized speech/language interventions with both adults and children involve complicated processes of assessment and goal-setting with the aim of achieving specific improvements. If these make no difference, and the same could be achieved by simply being warm, supportive, and encouraging, then we should not be spending time on them.

Generalized beneficial effects of being singled out for intervention can also operate via teachers. In a famous experiment, Rosenthal & Jacobson (1968) provided teachers with arbitrary information about which children in their class were likely to be "academic bloomers". They subsequently showed that the children so designated obtained higher intelligence test scores at a later point, even though they had started out no different from other children. In fact, despite its fame, the Rosenthal and Jackson result is based on data that, by modern standards, are far from compelling, and it is not clear just how big an effect can be produced by raising teacher expectations (Jussim & Harber, 2005). Nevertheless, the study provides strong motivation for researchers to be cautious about how they introduce interventions, especially in school settings. If one class is singled out to receive a special new intervention, it is likely that the teachers will be more engaged and enthusiastic than those who remain in other classes where it is "business as usual." We may then be misled into thinking that the intervention is effective, when in fact it is the boost to teacher engagement and motivation that has an impact.

Study participants themselves may also be influenced by the general positive impact of being in an intervention group – and indeed it has been argued that there can be an opposite effect – a **nocebo** effect – for those who know that they are in a control group, while others are receiving intervention. This is one reason why some studies are conducted as **double blind** trials – meaning that neither the participant nor the experimenter knows who is in which intervention group. But this is much easier to achieve when the intervention is a pill (when placebo pills can be designed to look the same as active pills) than when it involves communicative interaction between therapist and client. Consequently, in most studies in this field, those receiving intervention will be responding not just to the specific ingredients of the intervention, but also to any general beneficial effects of the therapeutic situation.

4.3 Identifying specific intervention effects by measures of mechanism

There are two approaches that can be taken to disentangle nonspecific effects from the specific impact of a particular intervention. First, we can include an active control group who get an equivalent amount of therapeutic attention, but directed towards different goals – equivalent to the placebo group in Figure 4.1. We will discuss this further in Section 6, which focuses on different approaches to control groups. The second is to include specially selected measures designed to clarify and highlight the active ingredients of the intervention. We will refer to these as **'mechanism' measures**, to distinguish them from outcome measures. This topic is covered more formally in Section 16.2, where the idea of mediators and moderators of intervention effects will be discussed. Here we give a couple of examples of how additional measures can be incorporated into a study to help distinguish specific intervention effects from general improvement.

Let's take the example of parent-based intervention with a late-talking toddler. In extended milieu therapy, the therapist encourages the parent to change their style of communication in naturalistic contexts in specific ways. The ultimate goal of the intervention is to enhance the child's language, but the mechanism is via changes in the parent's communicative style. If we were to find that the child's language improved relative to an untreated control group but there was no change in parental communication (our mechanism measure), then this would suggest we were seeing some general impact of the therapeutic contact, rather than the intended effect of the intervention.

To take another example, the theory behind the computerized Fast ForWord® (FFW) intervention maintains that children's language disorders are due to problems in auditory processing that lead them to be poor at distinguishing parts of the speech signal that are brief, non-salient, or rapidly changing. The intervention involves exercises designed to improve auditory discrimination of certain types of sounds, with the expectation that improved discrimination will lead to improved general language function. If, however, we were to see improved language without the corresponding change in auditory discrimination (a mechanism measure), this would suggest that the active ingredient in the treatment is not the one proposed by the theory.

Note that in both these cases it is possible that we might see changes in our mechanism measure, without corresponding improvement in language. Thus we could see the desired changes in parental communicative style with extended milieu therapy, or improved auditory discrimination with FFW, but little change in the primary outcome measure. This would support the theoretical explanation of the intervention, while nevertheless indicating it was not effective. The approach might then either be abandoned or modified – it could be that children would need longer exposure, for instance, to produce a clear effect.

The most compelling result, however, is when there is a clear difference between an intervention and a treated control group in both the mechanism measure and the outcome measure, with the two being positively related within the treated group. This would look like good evidence for a specific intervention effect that was not just due to placebo.

It is not always easy to identify a mechanism measure – this will depend on the nature of the intervention and how specific its goals are. For some highly specific therapies – e.g., a phonological therapy aimed at changing a specific phonological process, or a grammatical intervention that trains production of particular morphological forms, the mechanism measure might be similar to the kind of "near transfer" outcome measure that was discussed in Chapter 3 – i.e., a measure of change in performance on the particular skill that is targeted. As noted above, we might want to use a broader assessment for our main outcome measure, to indicate how far there is generalization beyond the specific skills that have been targeted.

4.4   Check your understanding

1. In their analysis of the original data on the Hawthorne effect, Levitt & List (2011) found that output rose sharply on Mondays, regardless of whether artificial light was altered. Should we be concerned about possible effects of the day of the week or the time of day on intervention studies? For instance: Would it matter if all participants were given a baseline assessment on Monday and an outcome asssessment on Friday? Or if those in the control group were tested in the afternoon, but those in the intervention group were tested in the morning?

2. EasyPeasy is an intervention for preschoolers which "provides game ideas to the parents of preschool children to encourage play-based learning at home, with the aim of developing children's language development and self-regulation". In a report by the Education Endowment Foundation, the rationale is given as follows: "The assumption within the theory of change is that the EasyPeasy intervention will change child self-regulation which will lead to accelerated development in language and communication and improved school readiness. The expectation is that this will be achieved through the nursery teachers engaging with the parents regarding EasyPeasy and the parents engaging with their children through the EasyPeasy games. As well as improved self-regulation and language and communication development from playing the games, the expectation is that there will also be an improved home learning environment due to greater parent-child interaction. The expected impact was that this will lead to an improvement in children's readiness to learn along with improved parental engagement with the school".

Suppose you had funds to evaluate EasyPeasy. What would you need to measure to test the theory of change? Which measures would be mechanism measures and which would be outcomes? What would you conclude if the outcomes improved but mechanism measures showed no change?

5

Limitations of the pre-post design: biases related to systematic change

5.1 Learning objectives

By the end of this chapter, you will be able to:

- Describe three limitations of common pre-post intervention designs and how these can be addressed.

At first glance, assessing an intervention seems easy. Having used the information in Chapter 3 to select appropriate measures, you administer these to a group of people before starting the intervention, and again after it is completed, and then look to see if there has been meaningful change. This is what we call a pre-post design, and it almost always gives a misleadingly rosy impression of how effective the intervention is. The limitations of such studies have been well-understood in medicine for decades, yet in other fields they persist, perhaps because the problems are not immediately obvious.

Figure 5.1 shows some real data from a study conducted by Bishop et al. (2006), in which children were trained in a computerized game that involved listening to sentences and moving objects on a computer screen to match what they heard. The training items assessed understanding of word order, in sentences such as "the book is above the cup", or "the girl is being kicked by the boy". There were two treatment conditions, but the difference between them is not important for the point we want to make, which is that there was substantial improvement on a language comprehension test (Test for Reception of Grammar, TROG-2) administered at baseline and again after the intervention in both groups. If we only had data from these two groups, then we might have been tempted to conclude that the intervention was effective. However, we also had data from a third, untrained group who just had "teaching as usual" in their classroom. They showed just as much improvement as the groups doing the intervention, indicating that whatever was causing the improvement, it wasn't the intervention.

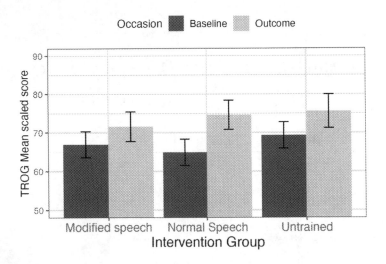

FIGURE 5.1
Mean comprehension test scores for three groups at baseline and outcome from a study by Bishop et al., 2006. Error bars shown standard errors.

To understand what was going on in this study, we need to recognize that there are several reasons why scores may improve after an intervention that are nothing to do with the intervention. These form the systematic biases that we mentioned in Chapter 1, and they can be divided into three kinds:

- Spontaneous improvement

- Practice effects

- Regression to the mean

5.2   Spontaneous improvement

We encountered spontaneous improvement in Chapter 1 where it provided a plausible reason for improved test scores in all three vignettes. People with acquired aphasia may continue to show recovery for months or years after the brain injury, regardless of any intervention. Some toddlers who are "late talkers" take off after a late start and catch up with their peers. And children in general get better at doing things as they get older. This is true not just for the more striking cases of late bloomers, but also for more serious and persistent problems. Even children with severe comprehension problems usually do develop more language as they grow older – it's just that the improvement may start from a very low baseline, so they fail to "catch up" with their peer group.

Most of the populations that speech and language therapists work with will show some spontaneous improvement that must be taken into account when evaluating intervention. Failure to recognize this is one of the factors that keeps snake-oil merchants in business: there are numerous unevidenced treatments offered for conditions like autism and dyslexia. Desperate parents, upset to see their children struggling, subscribe to these. Over time, the child's difficulties start to lessen, and this is attributed to the intervention, creating more "satisfied customers" who can then be used to advertise the intervention to others.

As shown in Figure 5.1, one corrective to this way of thinking is to include a control group who either get no intervention or "treatment as usual". If these cases do as well as the intervention group, we start to see that the intervention effect is not as it seems.

5.3 Practice effects

The results that featured in Figure 5.1 are hard to explain just in terms of spontaneous change. The children in this study had severe and persistent language problems for which they were receiving special education, and the time lag between initial and final testing was relatively short. So maturational change seemed an unlikely explanation for the improvement. However, practice effects were much more plausible.

A practice effect, as its name implies, is when you get better at doing a task simply because of prior exposure to the task. It doesn't mean you have to explicitly practice doing the task – rather it means that you can learn something useful about the task from previous exposure. A nice illustration of this is a balancing device that came as part of an exercise programme called Wii-fit. This connected with a controller box and the TV, so you could try specific exercises that were shown on the screen. Users were encouraged to start with exercises to estimate their "brain age". When the first author first tried the exercises, her brain age was estimated around 70 – at the time about 20 years more than her chronological age. But just a few days later, when she tried again, her brain age had improved enormously to be some 5 years younger than her chronological age. How could this be? She had barely begun to use the Wii-fit and so the change could not be attributed to the exercises. Rather, it was that familiarity with the evaluation tasks meant that she understood what she was supposed to do and could respond faster and apply strategies to optimize her performance.

It is often assumed that practice effects don't apply to psychometric tests, especially those with high test-retest reliability. However, that doesn't follow. When measured by a correlation coefficient, high reliability just tells you whether the rank ordering of a group of people will be similar from one test occasion to another – it does not say anything about whether the average performance will improve. In fact, we now know that some of our most reliable IQ tests show substantial practice effects persisting over many years e.g., (Rabbitt et al., 2004). One way of attempting to avoid practice effects is to use "parallel forms" of a test – that is different items with the same format. Yet that does not necessarily prevent practice effects, if these depend primarily on familiarization with task format and development of strategies.

There is a simple way to avoid confusing practice effects with genuine intervention effects, and it's the same solution as for spontaneous improvement – include a control group who don't receive the intervention. They should show exactly the same practice effects as the treated group, and so provide a comparison against which intervention-related change can be assessed.

5.4   Regression to the mean

Spontaneous improvement and practice effects are relatively easy to grasp intuitively using familiar examples. Regression to the mean is quite different – it appears counter-intuitive to most people, and it is frequently misunderstood.

It refers to the phenomenon whereby, if you select a group on the basis of poor scores on a measure, then the worse the performance is at the start, the greater the improvement you can expect on re-testing. The key to understanding regression to the mean is to recognize two conditions that are responsible for it: a) the same measure is used to select cases for the study and to assess their progress and b) the measure has imperfect test-retest reliability.

Perhaps the easiest way to get a grasp of what it entails is to suppose we had a group of 10 people and asked them each to throw a dice 10 times and total the score. Let's then divide them into the 5 people who got the lowest scores and the 5 who got the highest scores and repeat the experiment. What do we expect? Well, assuming we don't believe that anything other than chance affects scores (no supernatural forces or "winning streaks"), we'd expect the average score of the low-scorers to improve, and the average score of the high scorers to decline. This is because the most likely outcome for everyone is to get an average score on any one set of dice throws. So if you were below average on time 1, and average on time 2, your score has gone up; if you were above average on time 1, and average on time 2, your score has gone down.

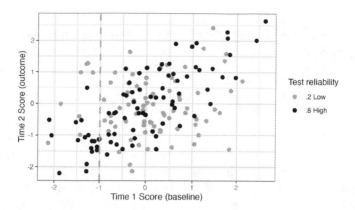

FIGURE 5.2

Simulated test scores for 160 children on tests varying in reliability. Individuals are colour-coded depending on whether they are given a test of low (.2) or high (.8) test-retest reliability. The simulations assume there is no effect of intervention, maturation, or practice

So that's the simplest case, when we have a score that is determined **only** by chance: i.e., the test-retest reliability of the dice score is zero.

Cognitive test scores are interesting here because they are typically thought of as comprising two parts: a "true" score, which reflects how good you really are at whatever is being measured, and an "error" score, which reflects random influences. Suppose, for instance, that you test a child's reading ability. In practice, the child is a very good reader, in the top 10% for her age, but the exact number of words she gets right will depend on all sorts of things: the particular words selected for the test (she may know "catacomb" but not "pharynx"), the mood she is in on the day, whether she is lucky or unlucky at guessing at words she is uncertain about. All these factors would be implicated in the "error" score, which is treated just like an additional chance factor or throw of the dice affecting a test score. A good test is mostly determined by the "true" score, with only a very small contribution of an "error" score, and we can identify it by the fact that children will be ranked by the test in a very similar way from one test occasion to the next; i.e., there will be good test-retest reliability. In other words, the correlation from time 1 to time 2 will be high.

Figure 5.2 shows simulated data for children seen on two occasions, when there is no effect of intervention, maturation, or practice. The vertical grey line divides the samples into those who score more than one SD below the mean at time 1. Insofar as chance contributes to their scores, then at time 2, we would expect the average score of such a group to improve, because chance pushes the group average towards the overall mean score. One can see by inspection that those given a low-reliability test tend to have average scores close to zero at time 2; those given a high-reliability test tend to remain below average. For children to the left of the grey vertical line, the mean time 1 scores for the two groups are very similar, but at time 2, the mean scores are 0.23 and -1.16 respectively for the low and high reliability tests.

The key point here is that if we select individuals on the basis of low scores on a test (as is often done, e.g., when identifying children with poor reading scores for a study of a dyslexia treatment), then, unless we have a highly reliable test with a negligible error term, the expectation is that the group's average score will improve on a second test session, for purely statistical reasons. In general, psychometric tests are designed to have reasonable reliability, but this varies from test to test and is seldom higher than .75-.8.

Cynically, one could say that if you want to persuade the world that you have an effective treatment, just do a study where you select poor performers on a test with low reliability! There is a very good chance you will see a big improvement in their scores on re-test, regardless of what you do in between.

So regression to the mean is a real issue in longitudinal and intervention studies. It is yet another reason why scores will change over time. X. Zhang & Tomblin (2003) noted that we can overcome this problem by using different tests to *select* cases for an intervention and to measure their outcomes. Or we can take into account regression to the mean by comparing our intervention group to a control group, who will be subject to the same regression to the mean.

5.5   Check your understanding

1.   Identify an intervention study on a topic of interest to you – you could do this by scanning through a journal or by typing relevant keywords into a database such as Google Scholar, Web of Science, or Scopus. If you are not sure how to do this, your librarian should be able to advise. It is important that the published article is available so you can read the detailed account. If an article is behind a paywall, you can usually obtain a copy by emailing the corresponding author.

   Your task is to evaluate the article in terms of how well it addresses the systematic biases covered in this chapter. Are the results likely to be affected by spontaneous improvement, practice effects, or regression to the mean? Does the study design control for these? Note that for this exercise you are not required to evaluate the statistical analysis: the answers to these questions depend just on how the study was designed.

2. A useful explainer for regression to the mean is found at this website: https://www.andifugard.info/regression-to-the-mean/(Fugard, 2023). It links to an app where you can play with settings of sample size, reliability, and sample selection criteria. Load the app and try to predict in advance how the results will be affected as you use the sliders to change the variables.

6

Estimating unwanted effects with a control group

6.1 Learning objectives

By the end of this chapter, you will be able to:

- Explain the importance of control groups;
- Understand the strengths and weaknesses of study designs using untreated or active controls.

The idea of a control group has already come up in Chapter 2, Chapter 4, and Chapter 5, where it has been noted that this provides one way of estimating how far an improvement from baseline to outcome is really due to the intervention. The logic is straightforward: if you have two groups – group A who has the intervention, and group B who does everything that group A does except for the intervention – then it should be possible to isolate the specific effect of the intervention. In this chapter we will discuss the different kinds of control groups that can be used. But first, it is important to address a question of ethics.

6.2 Is it ethical to include a control group?

The argument against a control group goes something like this: we think our intervention works, but we need to do a study to establish just how well it works. It would be unethical to withhold the intervention from one group of people because they would then be deprived of a potentially helpful treatment. This may seem particularly important if the client has serious problems – shouldn't we do all in our power to help?

This argument has as many holes in it as a Swiss cheese, and here's why. Do you already know if the treatment works? If the answer is Yes, then why are you doing a study? Presumably it is because you think and hope it works, but don't actually have evidence that it does. So you don't know and you need to find out. Freedman (1987) coined the term **clinical equipoise** to describe the situation when there is genuine uncertainty among practitioners about whether an intervention is better or worse than no intervention (or another intervention) – it is only ethical to do a trial of intervention when this is the case. But this can be hard to judge.

Unfortunately, the kinds of systematic bias discussed in Chapter 5 can lead practitioners to have an unwavering conviction that their intervention is effective: they see improvement in the clients they work with and assume it is because of what they have done. The harsh reality is that when properly controlled studies are conducted, it is often found that the intervention is not effective and that the improvement is just as great in controls as in the intervention group. There are numerous examples of this in mainstream medicine, as well as in education and allied health disciplines.

6.3   Possible adverse effects of intervention

Worse still, some interventions do more harm than good. The need to evaluate adverse impacts of intervention is well-recognized in mainstream medicine, where it is easy to see the increased risks of morbidity or mortality, or unpleasant side effects. In clinical trials, a new treatment or medication will pass through typically five stages or phases of trials. A each stage the number of participants is increased, and the early trials are specifically designed to test safety and efficacy, beginning with very small doses. Typically for drug studies, before human trials are attempted, treatments undergo substantial lab-based evaluation and sometimes animal testing.

In fields such as education or speech and language therapy it is typically assumed that interventions won't do any harm – and it is true that the types of adverse effect seen in medicine are not likely. Nevertheless, any intervention creates **opportunity costs** – the time (and sometimes money) that are spent in intervention can't be spent doing other things. In the study of computerized intervention by Bishop et al. (2006) head teachers were generally keen to take part, but they would ask what activities children should miss if they were to spend 15 minutes every day on our computerized tasks for a period of six weeks. Should we ask children to forfeit their breaks or to skip non-academic activities such as swimming? This would hardly be popular with the children, we felt. Or should they take time out of lessons in English or Maths?

Opportunity Costs in the Arrowsmith Program

The Arrowsmith Program is marketed as a neuroscience-based intervention for children with a wide range of specific learning disabilities. A review of the research literature on Arrowsmith finds no evidence of effectiveness, despite the program having been in existence since the 1980s (Bowen, 2020). One of the most comprehensive reviews of Arrowsmith research is in the D.Phil. thesis of Debra Kemp-Koo from the University of Saskatchewan (Kemp-Koo, 2013). In her introduction, Dr Kemp-Koo included an account of a study of children attending the private Arrowsmith school in Toronto: *All of the students in the study completed at least one year in the Arrowsmith program with most of them completing two years and some of them completing three years. At the end of the study many students had completed their Arrowsmith studies and left for other educational pursuits. The other students had not completed their Arrowsmith studies and continued at the Arrowsmith School.* **Most of the students who participated in the study were taking 6 forty minute modules of Arrowsmith programming a day with 1 forty minute period a day each of English and math at the Arrowsmith School. Some of the students took only Arrowsmith programming** *or took four modules of Arrowsmith programming with the other half of their day spent at the Arrowsmith school or another school in academic instruction (p. 34-35; our emphasis).* So children at Arrowsmith schools were spending one to three years working on unevidenced, repetitive exercises, rather than the regular academic curriculum. As Kemp-Koo (2013) remarked: *The Arrowsmith program itself does not focus on academic instruction, although some of these students did receive some academic instruction apart from their Arrowsmith programming.* **The length of time away from academic instruction could increase the amount of time needed to catch up with the academic instruction these students have missed.** *(p. 35; our emphasis).* It seems remarkable that children who have a range of specific learning difficulties are described as needing catch-up academic instruction because they have missed regular classroom activities because of time spent on the Arrowsmith program. This is a classic case of an intervention that carries a substantial opportunity cost – quite apart from the financial cost to families.

When we move to consider interventions outside the school setting, there are still time costs involved: an aphasic adult with poor mobility or a harassed mother with three small children who is reliant on public transport may have considerable difficulty attending clinic appointments. Of course, even if the intervention is ineffective, the clients may be glad to have the attention and professional interest of a therapist. But if that is something that is of importance to them, then that needs to be included in the outcome measures – i.e., it is necessary to demonstrate it is so, rather than just assuming this is the case.

For parent training programmes, there is an additional risk of making parents worry that they are inadequate, that they don't have time to devote to activities with their child, or that their child's difficulties are all their fault. Clearly, competent professionals will be at pains to counteract that perception, but it is an easy mindset to fall into (Weiss, 1991).

Finally, the person who is the focus of intervention may feel embarrassed, stressed, or stigmatized by being singled out for special treatment – this is most likely to be a potential problem for children who are pulled out of regular classes for intervention sessions. There is a poignant account of such an experience by the dyslexic chef Jamie Oliver who described how other children would sing "Special Needs, Special Needs" to the tune of *Let it Be* when he and another child were pulled out of class for individual tuition. Here again, a good professional will recognize that risk and do their best to address the concerns, but this needs to be thought about, rather than blithely assuming that because the intention of the therapist is good and helpful, their impact has to be beneficial.

All these potentially adverse impacts, including large investments of time, will generally be seen as a price well worth paying if the intervention is effective. At the end of the day, a decision whether to intervene or not should involve an appraisal of costs and benefits. Problems arise when costs are ignored and a rosy, but unevidenced, view of benefits is assumed.

6.4   Ethics of uncontrolled studies

As explained in Chapter 5, a study that just compares people on a language measure before and after intervention is generally uninterpretable, because there are numerous factors that could be responsible for change. Having a control group is not the only way forward: in Chapter 20, we discuss other types of research design that may be more appropriate for specific contexts. But in most situations where the aim is to test the effectiveness of an intervention for a particular condition, a study with a control group will be the best way to get useful information. Indeed, doing a study without controls is unethical if you end up investing the time of clients and professionals, and research funds, doing a study that cannot answer the question of whether the intervention is effective.

6.5 Treated vs. untreated controls

As noted above, a common design is to compare group A who receive intervention, with group B who are treated just the same but without any intervention. If you see no effect of intervention using this design, and the study is big enough to give adequate statistical power (see Chapter 13), then you can be pretty certain that the intervention is not effective. But if you do see a reliable improvement in group A, over and above and change in group B, can you be certain it was the intervention that made the difference?

There are two things to be concerned about here. First, in the previous chapter we discussed the kinds of very general impact that can be caused just by being in an intervention group. These work in the opposite way to the possible adverse effects discussed above: although some people may be embarrassed or stressed by being a target of intervention, others may be reassured or made more confident. A good therapist will do more than just administer an intervention in a standard fashion: they will form a relationship with their client which can itself be therapeutic. This, however, creates a confound if the control group, B, just has "business as usual" because it means we are really comparing group A, who gets a specific intervention plus a therapeutic relationship, with group B, who gets nothing.

A creative way of dealing with this confound is to ensure that group B does also get some kind of intervention, focusing on something that is worthwhile but different from the one that is the focus of the study. This is known as an **active control**. It is important to choose a comparison intervention that is not expected to have an impact on the main outcome measure – otherwise you could find yourself in a situation where everyone gets better but groups A and B don't differ, and you are left uncertain whether both interventions worked or neither did. On the other hand, it would be unethical to use an active control that was of no conceivable benefit. An example from the speech and language field is the study by Burgoyne et al. (2018). They compared preschoolers who had a parent-based language intervention with a group who had a parent-based intervention designed to improve motor skills. Another example comes from McGillion et al. (2017), who compared an intervention designed to increase caregiver contingent talk to infants with a control condition where caregivers viewed a video about healthy eating and tooth-brushing and were asked to spend 15 min a day with their child making these a habit.

The use of a contrasting intervention overcomes another possible concern about untreated controls, which is those in group B may actually do worse because they know they are missing out on an intervention that might be able to help them (see Chapter 4).

Another popular approach to controls is to have a two-stage study. Stage 1 is a classic controlled study where a treated group A is compared with an untreated group B. After the intervention phase, group B is then offered the intervention. At this point, group A may continue with the intervention, or intervention may be withdrawn. These kind of designs are called crossover or wait-list designs and will be discussed further in Chapter 19.

6.6 Check your understanding

Take one of the three vignettes from Chapter 1, or use an intervention method you are familiar with.

- Make a list of any possible harms you can think of that might be caused by the intervention. Remember that in behavioural areas, most harms are likely to be psychological, social, or financial.

- If you wanted to evaluate this intervention using an active control group (rather than an untreated control), what kind of intervention could you use for the controls?

7

Controlling for selection bias: randomized assignment to intervention

7.1 Learning objectives

By the end of this chapter, you will be able to:

- Explain why non-random allocation to intervention is problematic;
- Be aware of the differences between simple randomization, stratification, and minimization.

7.2 Randomization methods

In this chapter we consider how to select people for the experimental and control groups of an intervention study. This is a key element of a randomized controlled trial (RCT), which is widely regarded as a gold standard approach to the evaluation of interventions. The core idea is that once a target population has been identified for study, the experimenter allocates participants to intervention and control groups in a way that minimizes bias. This turns out to be much harder than you might imagine.

There is a big literature on methods of allocation to intervention. The main principle behind modern intervention studies is that allocation should be completely random, i.e., not predictable from any characteristics of participants. There are many approaches to treatment allocation that are designed to avoid bias but which are not random. We defer until later (Chapter 9) the question of what to do when people drop out of a trial. We assume we have a group of potential participants and have to decide who will be in the intervention group and who will be in the control group. Consider the following methods of allocation and note your judgement as to whether they are (a) reasonable ways of avoiding bias and (b) random.

A. Allocate people to the intervention group until the desired sample size is reached; then allocate all subsequent cases to the control group.

B. Allocate the more severe cases to the intervention group and the less severe to the control group.

C. Alternate between intervention and control: thus the first person on the list is an intervention case and the last one is a control.

D. For each potential participant, toss a coin and allocate those with 'heads' to the intervention group and those with 'tails' to the control group.

E. Create pairs of people matched on a key variable such as age, and then toss a coin to decide which member of each pair will be in the intervention group and which in the control group.

If you have read this far, it is hoped that you will identify that B is clearly a flawed approach. Not only does it fail to equate the intervention and control groups at the outset, but it also is likely to result in regression to the mean, so that greater gains will be seen for the intervention vs. the control group, for purely spurious statistical reasons. Despite its obvious flaws, this approach is sometimes seen in published literature – often taken to the extreme where an intervention group with a disorder is compared to a 'control' group with no disorder. This makes no sense at all – you are not controlling for anything unless the groups are equivalent at the outset and so cannot evaluate the intervention this way.

Approach A is an improvement, but it is still prone to bias, because there may be systematic differences between people who are identified early in the recruitment phase of a study and later on. For instance, those who sign up immediately may be the most enthusiastic. Approach A also creates problems for allocation concealment, which is discussed further in Chapter 10.

Most people think that Approach C as a reasonable way to achieve equal group sizes and avoid the kinds of "time of recruitment" issues that arise with Approach A. But this is not a random approach and would therefore be regarded as problematic by modern standards of trials design. One reason, which we will discuss at more length in Chapter 10, is that this method could allow the researcher to influence who gets into the study. Suppose the experimenter allocates aphasic patients to a study to improve word-finding according to some systematic method such as their serial order on a list, but notices that the next person destined for the intervention group has particularly profound difficulties with comprehension. It may be sorely tempting to ensure that he had an "odd" number rather than an "even" number and so ended up in the control group. Similar objections arise to methods such as using a person's case number or date of birth as the basis for intervention assignment: although this may be a fixed characteristic of the person and so appear to avoid bias, it raises the possibility that the experimenter could simply decide not to include someone in the study if they were destined for the intervention group and looked unpromising.

Approach D is clearly random, but it runs the risk that the two groups may be of unequal size and they may also be poorly matched on the pre-intervention measures, especially if sample sizes are small.

Approach E would seem like the logical solution: it avoids unequal sample sizes, ensures groups are comparable at the outset, yet includes randomization. This method does, however, have some disadvantages: it can be difficult to apply if one does not have information about the full sample in advance, and can be problematic if participants drop out of the trial, so the matching is disrupted.

We have seen the drawbacks of some of the approaches that might at first glance seem appropriate to avoid bias. Here we will briefly describe randomization methods that are used in contemporary medical trials, to give an idea of the range of possible approaches. Randomization is a fairly technical subject, and anyone designing a randomized controlled trial is advised to seek expert advice to ensure that the most effective method is used.

7.2.1   Simple randomization

In our earlier examples from the start of this chapter, we saw various examples of simple randomization (D-E). This takes the idea of the "coin toss" but adds in some degree of control of other variables which might affect the result. Typically, a coin is not used, as the random nature of the coin toss means that balanced numbers in each group will not be guaranteed. A randomization list is created by an individual who is unrelated to the trial. This list contains an equal number of allocations for a fixed number of participants which will have been pre-specified according to a sample size calculation.

In theory this might seem sufficient, but it ignores one aspect of real life: people will sign up for a trial but then drop out before assignment of intervention. This means that we potentially can have unbalanced groups despite following the list.

Furthermore as we saw earlier, simple randomization does not control key variables which may exert an effect on the outcome. For example, suppose we randomize children with a language difficulty into two groups, but we find that there are more boys than girls in one group. If we find an intervention effect as predicted, we can't be sure whether this is just because boys and girls respond differently.

Simple randomization is rarely (if at all) used in contemporary medical trials, but other approaches have been developed to counter the drawbacks encountered.

7.2.2 Random permuted blocks

The permuted blocks approach tries to overcome the problem of unbalanced groups by allocating individuals into blocks. The idea is that patients are allocated to intervention or control in blocks, so that at certain equally spaced points across the trial, equal numbers are ensured. Let's see how this works using an example:

Say we have chosen our block size to be four. So, in each block we have two cases assigned to intervention and two to the control group. The order of assignment of each block is random, for example, ABAB, AABB, ABBA, BBAA, BAAB, and so on. This means at any point in the trial, the randomization is only ever unbalanced by two individuals if the trial stops short or fails to recruit sufficient numbers of individuals, so the imbalance is minimal and statistical power is maintained.

However, the permuted blocks design has a downside; it is relatively easy for the researcher to see a pattern emerge in the allocation, so the randomization can become predictable. This means we then have to rely on the researcher to continue with correct allocation and not deviate, even if they have a strong belief in the intervention. It may seem remarkably cynical to assume that we need to take steps against researchers departing from random allocation to achieve a better result, but people do have unconscious biases that may lead them to do this, even if there is no conscious intent to deceive, and it is better to adopt an approach that makes it impossible for this to occur. Ideally, the randomization is done by someone other than the person running the trial.

7.2.3 Unequal randomization

Financial and practical constraints may limit the number of participants in a trial, raising the question of whether we might improve efficiency by having more people in the treatment than the control arm. The approach of unequal randomization allows for this: a fixed ratio is decided on before allocation of individuals begins, e.g., 2:1 intervention vs. control allocation. Use of this method should be considered carefully as there are both pros and cons. On the positive side, we may obtain a more accurate estimate of the effects of intervention, but we are prone to a reduction in power (Hey & Kimmelman, 2014); see also Chapter 13.

TABLE 7.1

Randomization with stratification by city and age band

	Intervention	Control
City 1, ages 5-7	23	24
City 2, ages 8-9	31	31
City 3, ages 10-11	28	27
City 1, ages 8-9	19	20
Cityl 2, ages 10-11	25	24
City 3, ages 5-7	35	35
City 1, ages 10-11	22	27
City 2, ages 5-7	24	23
City 3, ages 8-9	31	31

7.2.4   Stratification

We discussed above the possibility that randomized groups might differ in the proportion of males and females, making it difficult to disentangle the effect of the intervention from an effect of sex. There are usually other factors that could influence response to intervention: social and family background, presence of specific medical conditions, or educational attainment. If these are likely to be important, we can try to measure them in advance, but we cannot rule out the possibility that there may be other unknown variables that influence the treatment effect.

For characteristics that we can measure, stratification is one method that gives us some control, by splitting the allocation according to those variables. For example, suppose we have an intervention for school-aged children aged between 5 and 11 years. We recruit from three different cities: one with a high rate of deprivation, one with a good mix of social backgrounds, and one where most citizens are affluent home-owners. We would anticipate that there might be significant outcome differences in age and potentially differences between cities, so we use stratification to allow for this when we randomize children to intervention and control groups. The scheme would be as shown in Table 7.1.

7.2.5   Adaptive randomization

A problem with stratification is that it becomes weaker when the number of strata increases, particularly in small-scale trials. Suppose, for instance, we decided it was important to have six age-bands rather than three, and to separate boys and girls: we would soon find it difficult to achieve a balanced allocation.

TABLE 7.2
Example: randomized allocation by minimization

Prognostic factor	Intervention	Control
Sex		
Male	3	5
Female	5	3
Age band		
21-30	4	4
31-40	2	3
41-50	2	1
Risk factor		
High Risk	4	5
Low Risk	4	3

An alternative to stratification, which has become popular in recent years, is adaptive randomization (Hoare et al., 2013). For each new individual, the allocation to treatment ratio changes based on:

- the number of individuals already assigned to each treatment arm;
- the number of individuals in each stratification levels;
- the number of individuals in the stratum.

This method ensures that the allocation to groups remains balanced both overall and with any relevant stratifying variables or covariates, for example, gender or age.

7.2.6 Minimization

A specialized type of adaptive allocation called minimization is also very popular in clinical RCTs. Minimization was first proposed by Pocock & Simon (1975) and is a form of dynamic or adaptive randomization. Each participant's allocation is dependent on the characteristics and allocation of participants already allocated to groups. The idea is to minimize the imbalance by considering a range of prognostic factors for each allocation. This prevents the allocation of people with particular combinations of prognostic factors into one group. Table 7.2 illustrates this method. It uses an example from Scott et al. (2002) to show the situation after 16 individuals have been allocated to a trial using an example.

We now consider allocation of the 17th participant who is Male, aged 31-40, and High Risk. For each of these characteristics, we find the relevant row of the table and use Pocock & Simon (1975)'s method to work out the imbalance that would arise by adding this person to intervention or control, then summing over the imbalance for all the strata. We can then select the best allocation to minimize imbalance. In this case:

- If allocated to the intervention group, total imbalance is

$$|(3+1) - 5| + |(2+1) - 3| + |(4+1) - 5| = 1.$$

- If allocated to the control group, total imbalance is

$$|3 - (5+1)| + |2 - (3+1)| + |4 - (5+1)| = 7.$$

Since allocation to the intervention group gives imbalance of 1, whereas allocation to the control group would give imbalance of 7, this participant would be allocated to the intervention group.

7.3   Units of analysis: individuals vs. clusters

In the examples given above, allocation of intervention is done at the level of the individual. There are contexts, however, where it makes sense to use larger units containing groups of people such as classrooms, schools or clinics. Research in schools, in particular, may not readily lend itself to individualized methods, because that could involve different children in the same class receiving different interventions. This can make children more aware of which group they are in, and it can also lead to "spill-over" effects, whereby the intervention received by the first group affects other children who interact with them. To reduce spillover, it may make sense for children in classroom A to have one intervention and those in classroom B to have a control intervention. This is what happens in a clustered trial.

This method, however, raises new problems, because we now have a confound between classroom and intervention. Suppose the teacher in classroom A is a better teacher, or it just so happens that classroom B contains a higher proportion of disruptive pupils. In effect, we can no longer treat the individual as the unit of analysis. In addition, we may expect the children within a classroom to be more similar to one another than those in different classrooms, and this has an impact on the way in which the study needs to be analyzed. Clustered studies typically need bigger sample sizes than non-clustered ones. We will discuss this topic further in Chapter 18.

7.4 Check your understanding

Perform a literature search for an intervention/disorder of interest using the search term "randomized controlled trial" or "RCT", and download the article. (We are looking for a standard RCT, so do not pick a clustered RCT).

- Is the randomization process clearly described?

- Is it adequate?

- Did the researchers use stratification? If so, was the choice of variables for stratification sensible? If not, would stratification have helped?

8

The researcher as a source of bias

8.1 Learning objectives

By the end of this chapter, you will be able to:

- Explain the concept of researcher bias, and describe ways in which it can be counteracted;

- Appreciate how conflict of interest may lead researchers to report unduly rosy findings.

8.2 Motivations for researcher bias

The role of the researcher was already alluded to in Chapter 7, where it was noted that some kinds of randomization are problematic because they give the researcher wiggle room to omit or include participants. You might imagine that no honest person would distort their study this way, but it is all too easy to provide a post hoc rationalization for such decisions, or to be influenced by unconscious biases. Imagine you have planned a study of your favourite comprehension intervention, and you visit a school which is helping recruit children for the study. As you chat with the children in their classroom, you see that one little boy who met criteria for the study is highly overactive and seems unable to concentrate for more than a minute or two on anything. If you had control over intervention assignment, there might be a strong temptation to allocate him to the control group. Or you might rethink the study criteria and decide that you should exclude highly inattentive children. These kinds of reactions are understandable, but they create bias in the results.

8.3   Allocation concealment

This brings us to the topic of masking, or allocation concealment. Traditionally, the term "blinding" has been used, but it has been noted that this can be problematic when working with visually impaired people (Morris et al., 2007), so we will avoid it here. Nevertheless, you will hear the term in the context of the "double blind trial". This refers to the situation when neither the researcher nor the participants are aware of who is in the active or the control group. As noted in Chapter 4, it is not always feasible to keep people unaware of whether they are in an active intervention or control group, though use of active controls, as described in Chapter 6, may allow for this.

In allied health and educational fields, where resources are limited and the same person may be conducting the study and administering the intervention, particular care needs to be given to masking. Where feasible, we need to take steps to avoid what is termed **experimenter bias** in administration of baseline and outcome assessments.

8.4   The importance of masking for assessments

As discussed in Chapter 3, some assessments are more objective than others: it is relatively easy to be swayed by one's own desires and expectation when making a global rating of a person's communicative competence on a 4-point scale, much less so when administering a multiple choice comprehension test, where the task is to select a picture to match a spoken word. Nevertheless, there is evidence that, even with relatively objective tasks, researcher bias can creep in. Perhaps the clearest evidence comes from reviews that compare similar studies that are administered either with or without masking: where researchers are unaware of which group a subject is in, effects are typically smaller than when they are aware (and hence potentially able to bias the observations) (Holman et al., 2015).

Perhaps the most sobering example comes, not from an intervention study, but from data on a phonics screening test administered by teachers in UK schools in 2012. The score is supposed to indicate the number of items on a standard list of words and nonwords that a child reads accurately. Although there is some judgement required to tell whether a nonword is correctly pronounced, this should not exert a big effect on final scores, given that teachers were provided with a list of acceptable pronunciations. The results, however, which were published annually on a government website, showed clear evidence of scores being nudged up for some cases. We would expect a normal, bell-shaped

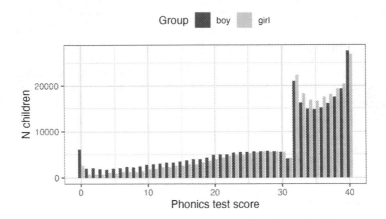

FIGURE 8.1
Numbers of children with specific scores on phonics screen. (Based on original data from Department for Education website, now deleted)

distribution of scores, but instead there was a sudden dip just below the pass mark and a corresponding excess of cases just about the pass mark. Since teachers were aware of the pass mark, they would have been able to nudge up the scores of children who were just one or two points below, and the distribution of scores is clear evidence that this happened. (Bishop, 2013a). A similar picture was observed in 2013, but the spike disappeared in 2014, when teachers were not told the pass mark in advance (Duff et al., 2015).

We don't usually think of teachers as dishonest or fraudulent. Some were opposed in principle to the phonics check and may have felt justified in not taking it seriously, but we suspect that most were doing their sincere best, but just did not think a slight tweak of a child's results would matter. There are anecdotal reports of teachers being unwilling to categorize children as "failing" at such a young age. And some may have been concerned that their personal reputation or that of their school might be damaged if too many children "failed". The possible reasons for nudging are many, and cannot be established post hoc, but the point we want to stress is that this kind of behaviour is not uncommon, often not done in a deliberately dishonest fashion, but is something that will happen if people are motivated to get one result or another.

In medical trials, the large amount of evidence for researcher bias on measures has led to a general consensus that outcome assessments should be done by someone who is unaware of who was in the intervention or control group. This is likely to add to the cost of conducting a trial, but it gives security that no nudging has taken place.

Another Questionable Classic Study

When researcher bias is discussed, we usually see citation to the work of Robert Rosenthal, who conducted a series of experiments in the 1960s, some of which have become classics in psychology. One of the most famous, which features in many undergraduate texts, is a study by Rosenthal & Fode (1963), which describes how student experimenters could be made to give biased evaluation of performance of rats in a maze-learning task. All the rats were from the same batch, but students were told they were either "maze bright" or "maze dull". Since there was no real difference between rats, they should have performed the same, but those designated as "maze bright" were faster in solving the maze than the "maze dull". Unfortunately, this classic study does not stand up to scrutiny. The numbers of students were small, and the results were not statistically robust when analyzed according to contemporary standards. Indeed, even at the time, Rosenthal's work came under strong criticism for methodological flaws (Barber & Silver, 1968). Nevertheless, Rosenthal's account of experimenter bias has been highly influential, forming part of the canon of "zombie psychological effects" that refuse to die, while Barber and Silver's critique has been largely forgotten. As with the Hawthorne effect, we see that few people return to look at the original data, and even if the research is debunked, it lives on because the story is just too perfect. Does this mean we do not need to worry about researcher effects? We recommend that it is good practice to take them into account when designing studies, but this story is further evidence of the need to look carefully at supposedly well-established findings.

8.5   Conflict of interest

Many medical journals, and increasingly journals in other fields, require authors to declare conflicts of interest, and a statement to this effect will be included with the published paper. Lo & Field (2009) define conflict of interest (COI) as:

> *"a set of circumstances that creates a risk that professional judgement or actions regarding a primary interest will be unduly influenced by a secondary interest."*

Typically, people think of COI as involving money. Someone who is marketing a computerized intervention, for instance, will be motivated to show it is effective – their business and livelihood would collapse if it was widely known to be useless. But COI – also sometimes referred to as "competing interests" – extends beyond the realms of finance. For example, a researcher's reputation may depend heavily on their invention of an intervention approach or there may be threats to cut government-funded services unless intervention is shown to be effective.

For those in vocational professions such as therapy or teaching, relatively poor working conditions may be endured in return for a sense of doing good. An intervention study with null results can be hard to accept if it means that the value of one's daily work is challenged.

In effect, most people involved in intervention studies want to show a positive effect for one reason or another. Unfortunately, there is evidence that conflict of interest leads to bias, with more positive results being reported when it is present (Friedman & Richter, 2004). It's not generally possible to avoid all conflict of interest, but the important thing is to recognize researcher bias as the rule rather than the exception, identify possible threats that this poses to study validity, take stringent steps to counteract these, and report openly any conflict of interest. We discussed above the ways in which results on outcome measures may be nudged up or down, often without any conscious attempt to mislead. But the impact of researcher bias can occur at all stages of an intervention study:

- At the stage of study design, the researcher may argue against including a control group – claiming ethical objections – because they are aware that it is much easier to show apparent intervention effects when there are no controls (see Chapter 5).

- In a controlled study, when allocating participants to intervention or control groups, the researcher may change inclusion criteria as the study progresses.

- Allocation to intervention or control groups may be under the control of a researcher who does not adopt truly random methods and so can determine who is in which group. Chapter 7 explains how randomization can overcome this.

- As noted above, if the person doing baseline and outcome assessments knows which intervention group a participant is in, then scores may be nudged.

This can be avoided by having assessments done by someone who is unaware of who got the intervention.

- When it comes to analysing the data, the researcher may decide on which variables to report, depending on the results. We discuss this problem and how to counteract it in Chapter 14.

- If the pattern of results does not show that the intervention is effective, then further checks and alternative analyses are conducted, whereas positive results are accepted without additional scrutiny.

- If the trial gives disappointing results, then the decision may be made not to report them. See Chapter 21 and Chapter 22 for discussion of this problem and suggestions for avoiding it.

The important point to recognize is that being a good scientist often conflicts with our natural human tendencies (Bishop, 2020). Scientists should be objective, and interpretation of results should not be swayed by personal likes and dislikes. On getting a positive intervention result, a good scientist will immediately ask: "Were there biases in my study that could have contributed to this finding?" – and indeed will not take offence if other scientists identify such factors. We need, of course, arguments in the opposite direction: a failure to see an intervention effect doesn't necessarily mean the intervention did not work – there could be aspects of study design that hide true effects, including too small a sample (see Chapter 13). At conferences involving allied health or education where people discuss intervention studies the question session is often dominated by people presenting arguments about why a null result may have missed a genuine effect, but it is much rarer to hear people questioning the validity of a positive result. It is a human tendency to accept information that fits with our desires and prejudices (in this case, that intervention is effective) and to reject contrary evidence. It also goes against our human nature to be critical of an intervention study conducted by a well-intentioned person who has put in a great deal of time and effort. But at the end of the day it is the quality of the evidence, rather than the feelings of researchers, that must take priority. Currently, we still know rather little about how best to help children and adults who struggle with speech, language, and communication. We will only change that if we take a rigorous and evidence-based approach to evaluating evidence.

8.6   Check your understanding

Once again, you need to find a published intervention study – this could be one you selected for a previous exercise or a new one.

- Does the published paper include a "conflict of interest" or "competing interests" statement?

- List the possible factors that might lead the experimenter to be biased in favour of finding a positive result.

- Consider the list of stages in the research process where experimenter bias could affect results: How have the researchers counteracted these?

9

Further potential for bias: volunteers, dropouts, and missing data

9.1 Learning objectives

By the end of this chapter, you will be able to:

- Distinguish potential sources of bias in who gets included in a study;
- Understand different approaches to handling missing data in an intervention study.

9.2 Who volunteers for research?

Many people who might benefit from interventions never take part in research evaluations. It has been noted that in psychological research, there is a bias for participants to be WEIRD, that is Western, educated, and from industrialized, rich, and democratic countries (Henrich et al., 2010). We are not aware of specific studies on participants in speech and language intervention studies, but it is likely that similar biases operate. It is important to know about the demographics of participants, as this may affect the extent to which results will be generalizable to the population at large. Researchers should report on the distribution of socio-economic, educational and cultural background of those taking part in intervention studies.

If the aim is to evaluate intervention targeted at disadvantaged groups, then it is important to think hard about how to optimize recruitment. Positive moves might include having a researcher or other intermediary who comes from a similar background to the target population, who will know the best ways to communicate and gain their confidence. Running a focus group may give insights into barriers to participation, which may be practical (finding time, or costs for transport or childcare) or social (feeling uncomfortable around people from a different social or ethnic background). Small things can make a big impact: we heard of one local community study with disadvantaged families that involved parent-training, which ended with parents being given a certificate on completing the training: the researchers said that this was appreciated by many parents, most of whom had never obtained any formal qualifications. Some of them went on to help recruit and train the next parent group.

The bottom line is that when planning to evaluate an intervention, it is important for researchers to think carefully about who the intervention is intended for. Once the target population is defined, then the aim should be to get as representative a sample as possible from that population. Involving the intended recipients of intervention in planning and running a study can both ensure that the intervention meets the needs of the community and that they are motivated to take part (see Graham et al. (2016)).

9.3   Dropouts after recruitment: intention to treat analysis

Marie is evaluating a phonological awareness training package with a group of "at risk" five-year-olds. She has adopted a randomized controlled design, with an active control group who get extra help with maths. She recruits 30 children to each group and runs the intervention for six weeks. However, two weeks in to the study, three children in the phonological awareness group and one from the control group drop out, and she learns that another child from the control group has been taken by his mother for intensive private speech and language therapy which includes phonological awareness training. Marie is left unsure what to do. Should she exclude these children from her study?

Every information sheet for an intervention study emphasizes that participants are free to drop out at any time, without giving a reason, with no adverse consequences. This is as it should be: it would not be ethical to coerce people to take part in research, and it would be foolish to try. Nevertheless, every researcher's heart sinks when this happens because it can really mess up a trial and make it harder to show an intervention effect.

If Marie were to consult a trials methodologist, she might be surprised at the answer. She is likely to be advised that she should not exclude participants who drop out or start other interventions. If people have withdrawn from the study, then their data cannot be included, but if there are children who are willing to be seen for the outcome assessments, their data should be included, even if they did not do the intervention they were assigned as planned. This is known as an **intention-to-treat** analysis – i.e., you conduct the analysis with people divided into groups according to how you intended to categorise them, regardless of what they actually did. At first glance, this seems crazy. We want to evaluate the intervention: surely we should not include people who don't get the full intervention! Even worse, the control group is supposed to contain people who didn't get the intervention – if we include some who obtained it by other means, this reduces the chances of seeing an effect.

The problems posed by drop-outs may be illustrated with an extreme example. Suppose Bridget Jones volunteers for a study of a miracle cure for obesity that involves giving up regular food for six weeks and consuming only a special banana milkshake that provides 500 calories per day. The control group consists of people who want to lose weight but are just given written materials explaining the calorie content of food. At the end of the six weeks, only 5 of the 50 original milkshake group are sticking to the diet, but they have all lost impressive amounts of weight compared to the control group, whose weight has gone down marginally. Bridget, together with 90% of the intervention group, found the diet impossible to stick to, got so hungry that she binged on chocolate, and actually gained weight. Is it reasonable to conclude that the milkshake diet worked? It depends on what question you are asking. If the question is "does extreme calorie restriction lead to weight loss?" the answer would appear to be yes. But that's not really what the study set out to do. It was designed to ask whether this specific intervention was effective for the kind of people enrolled in the study. Clearly, it wasn't – because most people found it impossible to stick to it – and it would be a bad idea to roll this out to larger samples of people.

In more realistic scenarios for allied health professionals, interventions are not likely to be so unacceptable to large numbers, but a key point is that drop-outs are seldom random. There are two opposite reasons why someone might give up on an intervention: they may just find it too hard to continue – and this could be because of difficulties in finding the time as much as difficulties with the intervention itself – or they might decide they are doing so well that they no longer need it. Either way, the group left behind after drop out is different from the one we started with.

It's instructive to look at the results from the little study by Bishop et al. (2006), who used computerized tasks to train comprehension in children (Chapter 5). Teachers and teaching assistants did the exercises with the children, three of whom dropped out. When we looked at the data, we found that the drop-outs had particularly low comprehension scores at the baseline. It is likely that these children just found the exercises too hard and became dispirited and unwilling to continue. Importantly, they were not a random subset of children. If we had excluded them from the analysis, we would no longer be comparing randomized groups who had been comparable at baseline – we would be comparing an intervention group that now omitted those with the worst problems with a control group that had no such omissions. This is known as a **per protocol** analysis, and it will overestimate the mean and underestimate the variance in the intervention effect and lead to biased results.

Nevertheless, if we only do an intention-to-treat analysis, then we lose an opportunity to learn more about the intervention. We would always complement it with an analysis of the drop-outs. Did they look different at baseline? Do we know why they dropped out? And what were their outcomes? For our comprehension study, this analysis gave a salutary message. The drop-outs showed substantial improvement on the outcome comprehension test, despite not having completed the intervention. This provides a clear example of how scores can change with practice on the test and regression to the mean (see Chapter 4).

9.4   Instrumental Variable analysis

A clever technique known as Instrumental Variable analysis can be used to obtain a more accurate estimate of an intervention effect when there are dropouts and/or control cases who received intervention, provided we can assess their outcomes. In effect, we do the analysis comparing those who did and did not actually do the intervention, but we adjust scores to take into account which group they were assigned to. This involves a two-step regression procedure, the details of which are beyond the scope of the current book. Instrumental variable analysis is illustrated by Huang (2018), who also gives a very clear explanation of the logic. The instrumental variable analysis gives a more accurate estimate of the intervention effect than the intention-to-treat analysis, without introducing the bias that occurs if we use a per protocol analysis.

Suppose we have run a trial that tests a particular intervention; we might find that not everyone conformed to the intervention, perhaps not attending all sessions. We can include attendance as a instrumental variable and get an adjusted estimate of our intervention effect on that basis (Figure 9.1). The important thing to keep in mind is that the instrumental variable is likely to be correlated with the intervention but not the outcome.

9.5   Missing data

People who drop out of trials are one source of missing data, but most trials will also have cases where just a subset of measures is missing. There are

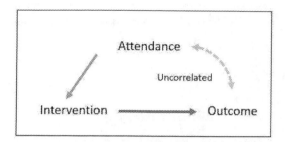

FIGURE 9.1
Path diagram of the instrumental variable approach.

numerous reasons for this: an assessment session may have over-run, test materials may have been lost, a computer may have crashed at a key moment, a researcher may have forgotten to record a result, or a participant may have been too tired, unwell, impatient, or anxious to continue. In the past, it was common for people to just drop missing data, but, just as with participant dropouts, this can create bias. In addition, some statistical methods require complete data for all participants, and it seems wasteful to drop a case when only part of the data is absent. Statisticians have devised a range of methods for handling missing data, typically using a method known as **imputation**, where missing data points are estimated from the existing data (Rioux & Little, 2021). How this is done depends on the underlying reason for what statisticians term "missingness". The easiest case to deal with is when it is completely random whether data are missing or not. The examples above where the missingness is caused by the researcher or equipment – a crashed computer, forgotten recording, or such – might feasibly be regarded as random. But where the missingness is down to something about the participant – for instance, refusal or inability to co-operate with an assessment – this is not random. And other situations may also be affected by the participant, even when this is less obvious. So if the session was not completed in available time, was this because the participant was slow to respond or distractible, or was it because of an external cause, such as a fire alarm?

Imputation typically involves estimating a predicted score from the available data. For instance, in the full dataset, an outcome measure may be predicted from baseline scores, age, and socio-economic status. We can use this information to create predicted outcomes for those who have missing data. This method, however, can create problems, because it gives the "best" estimate of the missing data, and will underestimate the variability of outcome scores. As explained by Van Buuren (2018), it is preferable to use an approach called **multiple imputation**, where several imputed datasets are created, each incorporating some random noise in the prediction. Statistical analysis is run for each imputed dataset, and the results are then pooled to give a final estimate

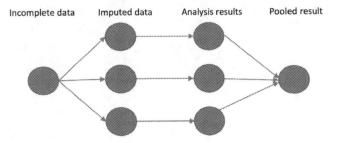

Incomplete data   Imputed data   Analysis results   Pooled result

FIGURE 9.2
Schematic stages of multiple imputation: Here 3 versions of the original dataset are generated with missing values imputed; the analysis is repeated for each version, and then the analysis results are pooled to give a single estimate of intervention effect

of the intervention effect. Figure 9.2 (based on an image from this website) shows the stages of processing. This approach is computationally demanding, especially for large datasets, but is a way of ensuring all the available data is used in an analysis, while minimizing bias. With modern computers it is tractable, and computer packages exist to handle the technical details.

What about the case where the missing data is not random? Then one really needs to think through the logic of each individual situation. For instance, suppose we have an assessment such as The Test of Word Reading Efficiency (Torgesen et al., 1999), where a pretest is used to familiarize the child with the task and check if they can read very simple words or nonwords. If the child can't read any items in the pretest, then the assessment is terminated. There is no standard score, but in this case, it is reasonable to assign a score at the test floor. Consider another case where children from one classroom have missing data on a reading assessment because the teacher did not wish to test the children. In such a case, the best option may be to use imputation, based on data from other assessments.

A useful overview of methods for handling missing data is provided in a website by Iris Eekhout: https://www.missingdata.nl/ (Eekhout, 2023).

9.6   Check your understanding

1. A study by Imhof et al. (2023) evaluated a video-coaching intervention with parents of children involved in a Head Start Program in the USA. Read the methods of the study and consider what measures were taken to encourage families to participate. How successful were

they? What impact is there on the study of: (a) failure to recruit sufficient families into the study, (b) drop-out by those who had been enrolled in the intervention?

2. Use of the flow chart shown in Figure 9.3 has become standard in medical trials, because it helps clarify the issues covered in this chapter. Find a study in the intervention literature which does not include a flow diagram, and see how much of the information in the flow chart can be completed by reading the Methods section. (The Imhof et al. study can also be used for this exercise).

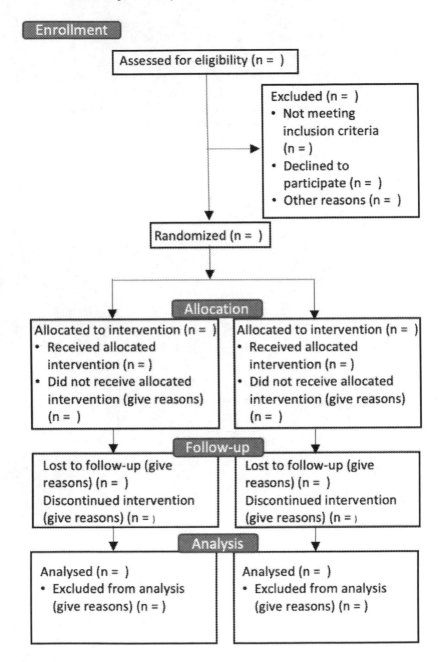

FIGURE 9.3
CONSORT flow diagram.

10

The randomized controlled trial as a method for controlling biases

10.1 Learning objectives

By the end of this chapter, you will be able to:

- Explain which biases from previous chapters are adequately controlled in a randomized controlled trial;

- Understand how reporting standards can help ensure study quality.

10.2 The ingredients of an RCT

The randomized controlled trial (RCT) is regarded by many as the gold standard method for evaluating interventions. In Chapter 15 we will discuss some of the limitations of this approach that can make it less than ideal for evaluating certain kinds of non-medical interventions. But in this chapter we'll look at the ingredients of an RCT that make it such a well-regarded method, and introduce the basic methods that can be used to analyze the results.

An RCT is effective simply because it is designed to counteract all of the systematic biases that were covered in previous chapters.

We cannot prevent changes in outcomes over time that arise for reasons other than intervention (Chapter 4), but with a control group we can estimate and correct for their influence. Noisy data in general arises either because of heterogenous participants, or because of unreliable measures: in a good RCT there will be strict selection criteria for participants and careful choice of outcome measures to be psychometrically sound (see Chapter 3). Randomization of participants to intervention and control groups avoids bias caused either by participants' self-selection of intervention group or researchers determining

TABLE 10.1
How the RCT design deals with threats to study validity

Biases	Remedies
Spontaneous improvement	Control group
Practice effects	Control group
Regression to the mean	Control group
Noisy data (1)	Strict selection criteria for participants
Noisy data (2)	Outcomes with low measurement error
Selection bias	Random assignment to intervention
Placebo effects	Participant unaware of assignment
Experimenter bias (1)	Experimenter unaware of assignment
Experimenter bias (2)	Strictly specified protocol
Biased drop-outs	Intention-to-treat/Instrumental variable analysis
Low power	A priori power analysis
False positives due to p-hacking	Registration of trial protocol

who gets what treatment. Also, as noted in Chapter 4 and Chapter 8, where feasible, both participants and experimenters are kept unaware of who is in which group, giving a **double blind** RCT.

10.3 Primary and secondary outcomes

In Chapter 14, we will explain problems caused by *p-hacking*, where many outcome measures are included but only the "significant" ones are reported. This problem has been recognized in the context of clinical trials for many years, which is why clinical trial protocols are usually registered specifying a primary outcome measure of interest: indeed, as is discussed further in Chapter 22, many journals will not accept a trial for publication unless it has been registered on a site such as https://clinicaltrials.gov/. Note, however, that, as will be discussed in Chapter 14, multiple outcomes may increase the statistical power of a study, and are not a problem if the statistical analysis handles the multiplicity correctly. Secondary outcome measures can also be specified, but reporting of analyses relating to these outcomes should make it clear that they are much more exploratory. In principle, this should limit the amount of data dredging for a result that is only loosely related to the anticipated intervention effect.

10.4   Reporting standards

RCTs have become such a bedrock of medical research that standards for reporting them have been developed. In Chapter 9 we saw the CONSORT flowchart, which is a useful way of documenting the flow of participants through a trial. CONSORT stands for Consolidated Standards of Reporting Trials, which are endorsed by many medical journals. Indeed, if you plan to publish an intervention study in one of those journals, you are likely to be required to show you have followed the guidelines. The relevant information is available on the 'Enhancing the QUAlity and Transparency Of health Research' (EQUATOR) network website (http://www.equator-network.org). The EQUATOR network site covers not only RCTs but also the full spectrum of guidelines of many types of clinical research designs.

For someone starting out planning a trial, it is worth reading the CONSORT Explanation and Elaboration document (Moher et al., 2010), which gives the rationale behind different aspects of the CONSORT guidelines. This may seem rather daunting to beginners, as it mentions more complex trial designs as well as a standard RCT comparing intervention and control groups, and it assumes a degree of statistical expertise. It is nevertheless worth studying, as adherence to CONSORT guidelines is seen as a marker of study quality, and it is much easier to conform to their recommendations if a study is planned with the guidelines in mind, rather than if the guidelines are only consulted after the study is done.

10.5   Check your understanding

1. The CONSORT Explanation and Elaboration document (Moher et al., 2010) notes the importance of specifying details of study design and implementation, including the following:

   - Eligibility criteria for participants,

   - Settings and locations where the data were collected,

   - The interventions for each group with sufficient details to allow replication, including how and when they were actually administered.

   Find a published intervention study in your area of interest, and compare reporting of these features with the description of what is required by Moher et al., 2010.

2. In Chapter 15 we consider drawbacks of the RCT design. Before you read that chapter, see if you can anticipate the issues that we consider in our evaluation. In what circumstances would an RCT be impractical or inappropriate?

11

The importance of variation

11.1 Learning objectives

By the end of this chapter, you will be able to:

- Appreciate why we need to measure variation as well as average effects of an intervention;

- Understand the terms "standard deviation" and "confidence interval".

A note of warning: Chapter 11 to Chapter 14 are the most challenging in the book, insofar as they introduce several statistical concepts that are frequently misunderstood. We have minimized the amount of mathematical material in these chapters and provide brief verbal summaries of the main points. Readers without a statistical background may find it helpful to try some of the exercises with interactive websites recommended in Check Your Understanding sections to gain a more intuitive understanding of the concepts covered here.

11.2 The importance of variability

When we are considering the impact of an intervention, we tend to focus on means: in particular on the difference in average outcomes between an intervention group and a control group. But what we are really interested in is variability – specifically, how much of the variability in people's outcomes can be attributed to the intervention, and how much can be regarded as random noise. The widely-used method of **analysis of variance** gives us exactly that information, by looking at the total variability in outcomes and considering what proportion is accounted for by the intervention.

To make this more concrete, consider the following scenario: You want to evaluate a weight loss program, Carbocut, which restricts carbohydrates. You compare 20 people on the Carbocut diet for 2 weeks with a control group where people are just told to stick to three meals a day (Program 3M). You find that

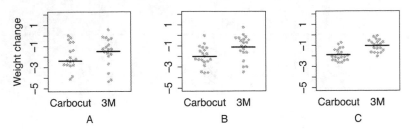

FIGURE 11.1
Simulated data from three studies comparing two diets. Each dot is the weight
change of an individual. The horizontal bars show the group mean. Overlap
between groups is greatest for study A and least for study C.

those on Carbocut have lost an average of 1 lb more weight than those on 3M.
Does that mean the diet worked? It's hard to say just from the information
provided: some people may think that the average loss is unimpressive; others
may think that it is encouraging, especially over such a short time scale. These
answers are about **real-life significance** of this amount of weight loss. But
they disregard another issue, the **statistical significance** of the weight loss.
That has nothing to do with meaningfulness in the real world, and everything
to do with **reproducibility** of the result. And this relates to the variability of
outcomes within each group.

11.2.1   Standard deviation

Consider Figure 11.1, which shows three different fictional studies. They have
the same sample size in each group, 20. The mean difference in lbs of weight
loss between the two diets (represented by the difference in the black horizontal
lines) is similar in each case, but the variation within each group (the spread
of points around the line) is greatest in scenario A, and least in scenario C.
The intervention effect is far less impressive in scenario A, where 7/20 people
in the Carbocut group score above the mean level of the 3M group, than in
group C, where only 2/20 score above the 3M mean.

The spread in scores within each group is typically measured by the **standard
deviation**, which is a mathematical way of representing how tightly bunched
a set of scores is around the mean. If we want to know how solid a result is, in
terms of how likely it would be to reproduce in a new study, we need to consider
not just the mean but also the standard deviation of the intervention effect.
This website (https://www.mathsisfun.com/data/standard-deviation.html)
explains how the standard deviation is computed, but in general, we can rely
on statistics programs to do the calculations for us.

Why does the standard deviation matter? This is because, as noted above, the
statistical significance of a finding, which indexes how likely a finding is to be

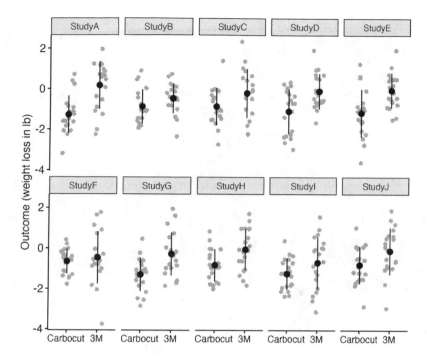

FIGURE 11.2

Simulated data from ten studies comparing two diets. The observed mean is shown as a black point and the SD as the black fins.

reproducible in future studies, considers all the variation in scores and looks at variation between groups *relative to the variation within each group*.

11.2.2 Confidence intervals

The **confidence interval** provides an index reflecting the precision of our estimate of the group difference. Suppose we were to run ten identical studies comparing the two diets, keeping the sample size the same each time. Figure 11.2 shows some simulated data of this scenario. In each case, the data were drawn by sampling from a population where the mean weight loss for Carbocut was 1 lb, and the mean loss for 3M was 0.25 lb. In both groups, the standard deviation was set to one. You may be surprised to see that the difference in means of the two groups fluctuates quite markedly, being substantial in study A and negligible in study F, with other studies intermediate. The same computer code was used to generate the three plots: the different results are just random fluctuations due to chance, which will occur when you use random number generators in simulations. This relates to the topic of **sampling**, which we will cover in Chapter 13. The main take-home message is that when we run a

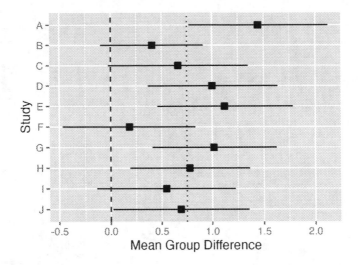

FIGURE 11.3
Mean difference (black square) with 95 per cent Confidence Intervals (fins) for
10 studies from previous Figure (11.2). Dashed line shows zero; dotted line
shows true population difference.

study and get a mean difference between groups, this is an **estimate** of the
true effect, which will contain some random noise.

When we report results from a study, we can report a **confidence interval**
(CI) around the estimated mean difference between groups, which gives an
indication of the uncertainty associated with the estimate; this will depend
on the standard deviation and the sample size, both of which feature in its
calculation. Figure 11.3 shows the mean group differences with 95% confidence
intervals for the 10 studies from Figure 11.2. Because these data were simulated,
we know that the true mean difference is 0.75, shown as a dotted vertical line,
and can see that the mean estimates from the ten studies generally cluster
around that value, but with a fair bit of variation. As we shall see in Chapter
13, high variation characterizes studies with small sample sizes: the confidence
intervals would be much smaller if we had 200 participants per group rather
than 20. The dashed vertical line denotes zero. If the confidence interval
includes zero, the study will fail to reach statistical significance on a t-test,
which is a method of analysis that we will explain in Chapter 12. We can see
that studies B, C, F, and I would all be non-significant on this criterion. We
will discuss in Chapter 13 how it is that we can get nonsignificant results even
if there is a true group difference – and how to avoid this situation.

Reporting of Confidence Intervals

Those who set standards for reporting clinical trial results generally encourage researchers to include confidence intervals as well as means. Reporting of confidence intervals is useful, not least because it encourages researchers and readers of research to appreciate that the results obtained from a study are just estimates of the true effect, and there may be a wide margin of uncertainty around results from small samples. Some have pushed back against reporting of confidence intervals, either because of conceptual objections and/or because the confidence interval is widely misunderstood. The correct interpretation of a 95 per cent confidence interval is that if you conducted your study over and over again, on 95 per cent of occasions the confidence interval that you get will contain the true population mean. Illustrating this point, in Figure 11.3, you can see that for most studies, the confidence interval contains the true value, .75, but there is one study, study A, where there is a very large difference between means and the confidence interval does not include the true value.

11.3   Check your understanding

1. Look at this website, https://shiny.rit.albany.edu/stat/confidence/, which has a simple interface that allows you to see how confidence intervals change as you vary the sample size and confidence level. First, check whether you understand what is shown on the plot. Can you predict what will happen if you:

    - Increase the number in the Sample Size box?

    - Change the Confidence Level?

    - Change the Standard Deviation?

    Try playing with the app by altering numbers in the boxes to see if your predictions are confirmed.

12

Analysis of a two-group RCT

12.1 Learning objectives

By the end of this chapter, you will be able to:

- Interpret output from a t-test;
- Understand why Analysis of Covariance is often recommended to analyze outcome data.

12.2 Planning the analysis

Statisticians often complain that researchers will come along with a set of data and ask for advice as to how to analyze it. In his Presidential Address to the First Indian Statistical Congress in 1938, Sir Ronald Fisher (one of the most famous statisticians of all time) commented:

> "To consult the statistician after an experiment is finished is often merely to ask him to conduct a post mortem examination. He can perhaps say what the experiment died of."

His point was that very often the statistician would have advised doing something different in the first place, had they been consulted at the outset. Once the data are collected, it may be too late to rescue the study from a fatal flaw.

Many of those who train as allied health professionals get rather limited statistical training. We suspect it is not common for them to have ready access to expert advice from a statistician. We have, therefore, a dilemma: many of those who have to administer interventions have not been given the statistical training that is needed to evaluate their effectiveness.

We do not propose to try to turn readers of this book into expert statisticians, but we hope to instill a basic understanding of some key principles that will make it easier to read and interpret the research literature, and to have fruitful discussions with a statistician if you are planning a study.

12.3   Understanding p-values

When we do an intervention study, we want to find out whether a given intervention works. Most studies use an approach known as **Null Hypothesis Significance Testing**, which gives us a rather roundabout answer to the question. Typically, findings are evaluated in terms of p-values, which tell us *what is the probability (p) that our result, or a more extreme one, could have arisen if there is no real effect of the intervention* – or in statistical jargon, if the **null hypothesis** is true. The reason why we will sometimes find an apparent benefit of intervention, even when there is none, is down to random variation, as discussed in Chapter 1. Suppose in a hypothetical world we have a totally ineffective drug, and we do 100 studies in large samples to see if it works. On average, in five of those studies (i.e., 5 per cent of them), the p-value would be below .05. And in one study (1 per cent), it would be below .01. The way that p-values are calculated assumes certain things hold true about the nature of the data ("model assumptions"): we will say more about this later on.

P-values are very often misunderstood, and there are plenty of instances of wrong definitions being given even in textbooks. The p-value is *the probability of the observed data or more extreme data, if the null hypothesis is true.* It does *not* tell us the probability of the null hypothesis being true. And it tells us nothing about the plausibility of the alternative hypothesis, i.e., that the intervention works.

An analogy might help here. Suppose you are a jet-setting traveller and you wake up one morning confused about where you are. You wonder if you are in Rio de Janiero – think of that as the null hypothesis. You look out of the window, and it is snowing. You decide that it is very unlikely that you are in Rio. You reject the null hypothesis. But it's not clear where you are. Of course, if you knew for sure that you were either in Reykjavík or Rio, then you could be pretty sure you were in Reykjavík. But suppose it was *not* snowing. This would not leave you much the wiser.

A mistake often made when interpreting p-values is that people think it tells us something about the probability of a hypothesis being true. That is not the case. There are alternative Bayesian methods that can be used to judge the relatively likelihood of one hypothesis versus another, given some data, but they do not involve p-values.

A low p-value allows us to reject the null hypothesis with a certain degree of confidence, but this does no more than indicate "something is probably going on here – it's not just random" – or, in the analogy above, "I'm probably not in Rio."

TABLE 12.1
Possible outcomes of hypothesis test

	Ground truth	
	Intervention effective	Intervention ineffective
Reject Null Hypothesis	True Positive	False Positive **(Type I error)**
Do Not Reject Null Hypothesis	False Negative **(Type II error)**	True Negative

Criticisms of the Use of p-values

There are many criticisms of the use of p-values in science, and a good case can be made for using alternative approaches, notably methods based on Bayes theorem. Our focus here is on Null Hypothesis Significance Testing in part because such a high proportion of studies in the literature use this approach, and it is important to understand p-values in order to evaluate the literature. It has also been argued that p-values are useful provided people understand what they really mean (Lakens, 2021).

One reason why many people object to the use of p-values is that they are typically used to make a binary decision: we either accept or reject the null hypothesis, depending on whether the p-value is less than a certain value. In practice, evidence is graded, and it can be more meaningful to express results in terms of the amount of confidence in alternative interpretations, rather than as a single accept/reject cutoff (Quintana & Williams, 2018).

In practice, p-values are typically used to divide results into "statistically significant" or "non-significant", depending on whether the p-value is low enough to reject the null hypothesis. We do not defend this practice, which can lead to an all-consuming focus on whether p-values are above or below a cutoff, rather than considering effect sizes and strength of evidence. However, it is important to appreciate how the cutoff approach leads to experimental results falling into 4 possible categories, as shown in Table 12.1.

The ground truth is the result that we would obtain if we were able to administer the intervention to the whole population – this is of course impossible, but we assume that there is some general truth that we are aiming to discover by running our study on a sample from the population. We can see that if the intervention really is effective and the evidence leads us to reject the null hypothesis, we have a True Positive, and if the intervention is ineffective and we accept the null hypothesis, we have a True Negative. Our goal is to design our study so as to maximize the chances that our conclusion will be correct, but there two types of outcome that we can never avoid, but which we try to minimize, known as Type I and Type II errors. We will cover Type II errors in Chapter 13 and Type I in Chapter 14.

12.4 What kind of analysis is appropriate?

The answer to the question, "How should I analyze my data?" depends crucially on what hypothesis is being tested. In the case of an intervention trial, the hypothesis will usually be "Did intervention X improve the outcome Y in people with condition Z?" There is, in this case, a clear null hypothesis – that the intervention was ineffective, and the outcome of the intervention group would have been just the same if it had not been done. The null hypothesis significance testing approach answers just that question: it tells you how likely your data are if the the null hypothesis were true. To do that, you compare the distribution of outcome scores in the intervention group and the control group. And as emphasized earlier, we don't just look at the difference in mean outcomes between two groups; we consider whether that difference is greater than you'd expect given the variation *within* the two groups. (This is what the term "analysis of variance" refers to).

If the observed data indicate a difference between groups that is unlikely to be due to chance, this could be because the intervention group does better than the control group or because they do worse than the control group. If you predict that the two groups might differ but you are not sure whether the control group or the intervention group will be superior, then it is appropriate to do a **two-tailed test** that considers the probability of both outcomes. If, however, you predict that any difference is likely to be in favour of the intervention group, you can do a **one-tailed test**, which focuses only on differences in the predicted direction.

12.5 Sample dataset with illustrative analysis

To illustrate data analysis, we will use a real dataset that can be retrieved from the ESRC data archive (Burgoyne et al., 2016). We will focus only on a small subset of the data, which comes from an intervention study in which teaching assistants administered an individual reading and language intervention to children with Down syndrome. A wait-list RCT design was used (see Chapter 19), but we will focus here on just the first two phases, in which half the children were randomly assigned to intervention and the remainder formed a

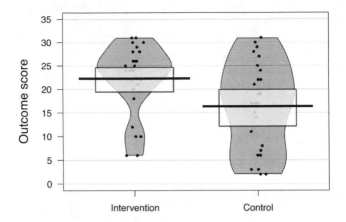

FIGURE 12.1

Data from RCT on language/reading intervention for Down syndrome by Burgoyne et al., 2012.

control group. Several language and reading measures were included in the study, giving a total of 11 outcomes. Here we will illustrate the analysis with just one of the outcomes – letter-sound coding – which was administered at baseline (t1) and immediately after the intervention (t2). Results from the full study have been reported by Burgoyne et al. (2012).

Figure 12.1 shows results on letter-sound coding after one group had received the intervention. This test had also been administered at baseline, but we will focus first just on the outcome results.

Raw data should always be inspected prior to any data analysis, in order to just check that the distribution of scores looks sensible. One hears of horror stories where, for instance, an error code of 999 got included in an analysis, distorting all the statistics. Or where an outcome score was entered as 10 rather than 100. Visualizing the data is useful when checking whether the results are in the right numerical range for the particular outcome measure. There are many different ways of visualizing data: it is best to use one that shows the distribution of data, and even individual data points if the sample is not too big. S. Zhang et al. (2022) showed that the traditional way of presenting only means and error bars in figures can be misleading and can lead people to overestimate group differences. A pirate plot such as that in 12.1 is one useful way of showing means and distributions as well as individual data points (Phillips, 2017). A related step is to check whether the distribution of the data meets the assumptions of the proposed statistical analysis. Many common statistical procedures assume that data are normally distributed on an interval scale (see Chapter 3). Statistical approaches to

TABLE 12.2

Summary statistics on letter-sound coding outcomes after intervention

Group	N	Mean	SD
Intervention	28	22.286	7.282
Control	26	16.346	9.423

checking of assumptions are beyond the scope of this book, but there are good sources of information on the web, such as this website for linear regression: http://www.sthda.com/english/articles/39-regression-model-diagnostics/161-linear-regression-assumptions-and-diagnostics-in-r-essentials/. But just eye-balling the data is useful and can detect obvious cases of non-normality, cases of ceiling or floor effects, or "clumpy" data, where only certain values are possible. Data with these features may need special treatment, and it is worth consulting a statistician if they apply to your data. For the data in Figure 12.1, although neither distribution has a classically normal distribution, we do not see major problems with ceiling or floor effects, and there is a reasonable spread of scores in both groups.

The next step is just to compute some basic statistics to get a feel for the effect size. Table 12.2 shows the mean and standard deviation on the outcome measure for each group. The mean is the average of the individual datapoints shown in Figure 12.1, obtained by just summing all scores and dividing by the number of cases. The standard deviation gives an indication of the spread of scores around the mean, and as we have seen, is a key statistic for measuring an intervention effect. In these results, one mean is higher than the other, but there is overlap between the groups. Statistical analysis gives us a way of quantifying how much confidence we can place in the group difference: in particular, how likely is it that there is no real impact of intervention and the observed results just reflect the play of chance. In this case we can see that the difference between means is around 6 points, and the average SD is around 8.

12.5.1 Simple t-test on outcome data

The simplest way of measuring the intervention effect is to just compare outcome measures on a t-test. We can use a one-tailed test with confidence, assuming that we anticipate outcomes will be better after intervention. One-tailed tests are often treated with suspicion, because they can be used by researchers engaged in p-hacking (see Chapter 14), but where we predict a directional effect, they are entirely appropriate and give greater power than a two-tailed test: see

TABLE 12.3
T-test on outcomes

t	df	p	mean diff.	lowerCI	upperCI
2.602	52	0.006	5.94	2.117	Inf

this blogpost by Daniël Lakens: http://daniellakens.blogspot.com/2016/03/one-sided-tests-efficient-and-underused.html(Lakens, 2016).

When reporting the result of a t-test, researchers should always report all the statistics: the value of t, the degrees of freedom, the means and SDs, and the confidence interval around the mean difference, as well as the p-value. This not only helps readers understand the magnitude and reliability of the effect of interest: it also allows for the study to readily be incorporated in a meta-analysis. Results from a t-test for the data in Table 12.2 are shown in Table 12.3. Note that with a one-tailed test, the confidence interval on one side will extend to infinity (denoted as Inf): this is because a one-tailed test assumes that the true result is greater than a specified mean value and disregards results that go in the opposite direction.

12.5.2   T-test on difference scores

The t-test on outcomes is easy to do, but it misses an opportunity to control for one unwanted source of variation, namely individual differences in the initial level of the language measure. For this reason, researchers often prefer to take difference scores: the difference between outcome and baseline measures, and apply a t-test to these. In fact, that is what we did in the fictional examples of dietary intervention above: we reported *weight loss* rather than weight for each group before and after intervention. That seems sensible, as it means we can ignore individual differences in initial weights in the analysis.

While this had some advantages over reliance on raw outcome measures, it also has disadvantages, because the amount of change that is possible from baseline to outcome is not the same for everyone. A child with a very low score at baseline has more "room for improvement" than one who has an average score. For this reason, analysis of difference scores is not generally recommended.

12.5.3   Analysis of covariance on outcome scores

Rather than taking difference scores, it is preferable to analyze differences in outcome measures after making a statistical adjustment that takes into account the initial baseline scores, using a method known as analysis of covariance or ANCOVA. In practice, this method usually gives results that are similar to those you would obtain from an analysis of difference scores, but the precision is greater, making it easier to detect a true effect. However, the data do need

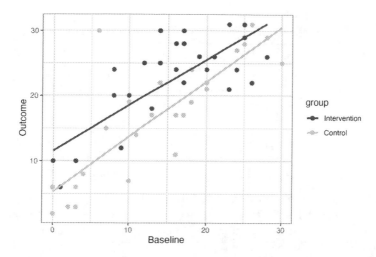

FIGURE 12.2
Baseline vs. outcome scores in the Down syndrome data.

TABLE 12.4
Analysis of outcome only (ANOVA)

Effect	DFn	DFd	F	p	ges
group	1	52	6.773	0.012	0.115

to meet certain assumptions of the method. This website walks through the steps for performing an ANCOVA in R, starting with a plot to check that there is a linear relationship between pretest vs. posttest scores in both groups – i.e., the points cluster around a straight line, as shown in Figure 12.2.

Inspection of the plot confirms that the relationship between pretest and outcome looks reasonably linear in both groups. Note that it also shows that there are rather more children with very low scores at pretest in the control group. This is just a chance finding – the kind of thing that can easily happen when you have relatively small numbers of children randomly assigned to groups.

Table 12.4 shows the Down syndrome data analyzed using ANOVA to compare only the outcome scores, and Table 12.5 shows the same data analyzed using ANCOVA to adjust scores for the baseline values.

The effect size is shown as *ges*, which stands for "generalized eta squared." You can see there is a large *ges* value, and correspondingly low p-value for the baseline term, reflecting the strong correlation between baseline and outcome shown in Figure 12.2. In effect, with ANCOVA, we adjust scores to remove

TABLE 12.5
Outcome adjusted for pretest scores (ANCOVA)

Effect	DFn	DFd	F	p	ges
baseline	1	51	94.313	0.000	0.649
group	1	51	9.301	0.004	0.154

the effect of the baseline on the outcome scores; in this case, we can then see a slightly stronger effect of the intervention: the effect size for the group term is higher and the p-value is lower than with the previous ANOVA.

For readers who are not accustomed to interpreting statistical output, the main take-away message here is that you get a better estimate of the intervention effect if the analysis uses a statistical adjustment that takes into account the baseline scores.

> **Analysis of Variance and Linear Regression**
>
> Mathematically, the t-test is equivalent to two other methods: analysis of variance and linear regression. When we have just two groups, all of these methods achieve the same thing: they divide up the variance in the data into variance associated with group identity and other (residual) variance, and provide a statistic that reflects the ratio between these two sources of variance. This is illustrated with the real data analyzed in Table 12.4. The analysis gives results that are equivalent to the t-test in Table 12.3: if you square the t-value, it is the same as the F-test. In this case, the p-value from the t-test is half the value of that in the ANOVA: this is because we specified a one-tailed test for the t-test. The p-value would be identical to that from ANOVA if a two-tailed t-test had been used.

12.5.4 Non-normal data

We've emphasised that the statistical methods reviewed so far assume that data will be reasonably normally distributed. This raises the question of what to do if we have results that don't meet that assumption. There are two answers that are typically considered. The first is to use **non-parametric statistics**, which do not require normally-distributed data (Corder & Foreman, 2014). For instance, the **Mann-Whitney U test** can be used in place of a t-test. An alternative approach is to **transform** the data to make it more normal. For instance, if the data is skewed with a long tail of extreme values, a log transform may make the distribution more normal. This is often appropriate if, for instance, measuring response times. Many people are hesitant about transforming data – they may feel it is wrong to manipulate scores in this way. There are two answers to that. First, it does not make sense to be comfortable with nonparametric statistics and uncomfortable with data transformation, because nonparametric methods involve transformation – typically they work by rank ordering data, and then working with the ranks, rather than the original scores. Second, there is nothing inherently correct about measuring things on a linear scale: scientists often use log scales to measure quantities that span a wide range, such as the Richter scale for earthquakes or the decibel scale for sound volume. So, provided the transformation is justified as helping make the data more suited for statistical analysis, it is a reasonable thing to do. Of course, it is *not* appropriate to transform data just to try and nudge data into significance. Nor is it a good idea to just wade in and apply a data transformation without first looking at the data to try to understand causes of non-normality. For instance, if your results showed that 90% of people had no effect of intervention, and 10% had a huge effect, it might make more sense to consider a different approach to analysis altogether, rather than shoe-horning results into a more normal distribution. We will not go into details of nonparametric tests and transformations here, as there are plenty of good sources of information available. We recommend consulting a statistician if data seem unsuitable for standard parametric analysis.

12.5.5 Linear mixed models (LMM) approach

Increasingly, reports of RCTs are using more sophisticated and flexible methods of analysis that can, for instance, cope with datasets that have missing data, or where distributions of scores are non-normal.

An advantage of the LMM approach is that it can be extended in many ways to give appropriate estimates of intervention effects in more complex trial designs – some of which are covered in Chapter 17 to Chapter 20. Disadvantages of this approach are that it is easy to make mistakes in specifying the model for analysis if you lack statistical expertise, and the output is harder for non-specialists to understand. If you have a simple design, such as that illustrated in this chapter, with normally distributed data, a basic analysis of covariance is perfectly adequate (O'Connell et al., 2017).

Table 12.6 summarizes the pros and cons of different analytic approaches.

12.6 Check your understanding

1. Figure 12.3 shows data from a small intervention trial in the form of a boxplot. If you are not sure what this represents, try Googling for more information. Check your understanding of the annotation in the top left of the plot. How can you interpret the p-value? What does the 95% CI refer to? Is the "mean.diff" value the same as shown in the plot?

TABLE 12.6
Analytic methods for comparing outcomes in intervention vs. control groups.

Method	Features	Ease of understanding	Flexibility
t-test	Good power with 1-tailed test. Suboptimal control for baseline. Assumes normality.	High	Low
ANOVA	With two-groups, equivalent to t-test, but two-tailed only. Can extend to more than two groups.	...	...
Linear regression/ ANCOVA	Similar to ANOVA, but can adjust outcomes for covariates, including baseline.	...	...
LMM	Flexible in cases with missing data, non-normal distributions.	Low	High

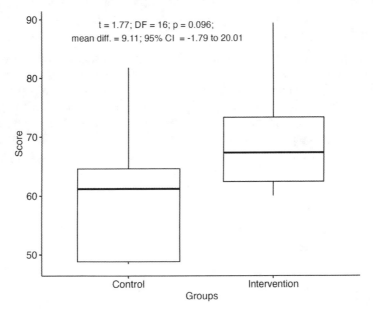

FIGURE 12.3
Sample data for t-test in boxplot format.

2. Find an intervention study of interest and check which analysis method was used to estimate the intervention effect. Did the researchers provide enough information to give an idea of the effect size or merely report p-values? Did the analysis method take into account baseline scores in an appropriate way?

13

How big a sample do I need?
Statistical power and type II errors

13.1 Learning objectives

By the end of this chapter, you will be able to:

- Appreciate how studies with small samples may lead us to conclude wrongly that an intervention is ineffective;

- Understand how effect size is measured and how it relates to statistical power;

- Explain what a type II error is, and how it may arise when sample size is too small.

13.2 Sampling

We start with further exploration of our fictitious diet comparison of Carbocut vs. 3M. Suppose there are two people who tried Carbocut and lost an average of 4 lb over 2 weeks. Another two people tried Program 3M and lost an average of 1 lb over 2 weeks. Would that convince you that Carbocut is superior? What if we tell you that one person in the Carbocut group lost 6 lb and the other gained 2 lb; in the Program 3M group, one person lost 5 lb and one gained 3 lb. Few people would find this convincing evidence that Carbocut is superior. With only two people per group, the average is quite unstable and highly dependent on the specific people in the group. Things would look more convincing if you had 1000 people in each group and still saw a 4 lb loss vs. a 2 lb loss. Individual weights would fluctuate, but the average is a much better indication of what is typical when we have a larger sample, because individual people with extreme scores will have less impact.

How could we know the real truth of just how much weight loss is associated with each diet? In theory, there is a way we could find out – we could recruit every single person in the world who meets some selection criterion for inclusion in the study and randomly assign them to one or the other diet, and then measure weight loss. Of course, in practice, that is neither possible nor ethical. Rather, the aim of research is to test a group of people whose results can be generalized to the broader population. So we recruit a **sample** which is intended to represent that much larger population. As the example above demonstrates, we need to think about sample size, as this will affect how much confidence we can place in the estimates from the sample. We know that two people is too small a sample. 1000 seems much better but, outside of medical trials, is often too expensive and difficult to achieve. So how do we decide the optimal sample?

13.3 Effect size

The *standardized effect size* is a way of describing the difference between groups that takes into account the standard deviation. There are various ways of measuring this: a simple measure is **Cohen's d**, which is the difference between groups measured in **standard deviation** units. Cohen's d is computed by dividing the mean difference between groups by the pooled standard deviation. For conditions A, B, and C in Figure 11.1, Cohen's d is .4, .8, and 1.5, respectively.

If we know what Cohen's d is, this gives us a sense of how much overlap there is between groups. Figure 13.1 shows the relationship between Cohen's d and overlap between groups as density plots, where the height of the distribution represents the frequency of individuals with that score: the larger the effect size, the less the overlap. It can be sobering to note that for most effective educational and therapeutic interventions, effect sizes are seldom greater than .3 or .4 (Kraft, 2023). Thus there is a lot of overlap, and in effect we have to pick out a meaningful signal – an intervention effect – from a very noisy background.

13.4 Sample size affects accuracy of estimates

Once we have an idea of the likely effect size in our study, we can estimate how big a sample we need. If we have a small effect size, then we need many observations to get a reliable estimate of the true effect of intervention. Previously we have shown plots where individual points correspond to scores

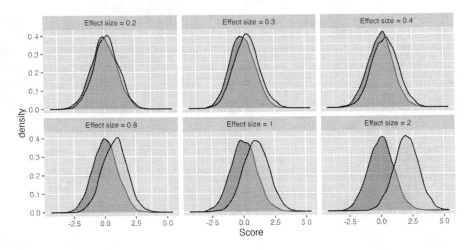

FIGURE 13.1
Overlap in distributions of scores of two groups for different effect sizes
(Cohen's d).

of individuals. In Figure 13.2 there is an important difference: each blob here
represents the *observed mean* from a sample of a given size, taken from a
population where the true effect size, shown as a dotted horizontal line, is 0.3.

Notice how the observed means jump around from sample to sample when
sample size is small but become more stable once we have 80 or more participants per group. When there are 320 participants per group, the means are
estimated pretty accurately, and the dots bunch around the mean line for the
group, but with 10 per group, it is very unreliable, and in some of our studies
the mean of the black group is lower than the mean of the grey group, which
is opposite to what is true in the population.

13.5   Type II error

A Type II error is the same as a **false negative**. It is the error that occurs
when the null hypothesis is not rejected but a true effect is actually present. In
other words, the data lead us to conclude an intervention doesn't work when
it really does have an effect.

Suppose a researcher wants to test a well-established and frequently-replicated
result: children whose parents read to them more frequently obtain better
results when they are tested on vocabulary. Twenty families are split into two
groups; in the first group, parents are encouraged to read with their child each

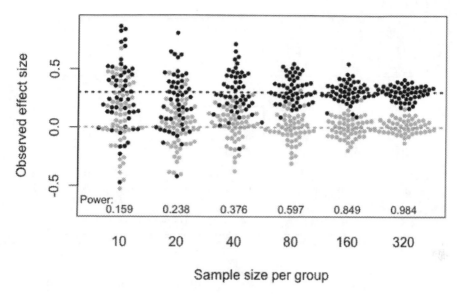

FIGURE 13.2
Simulated mean scores from samples of varying size, drawn from populations with either a null effect (grey) or a true effect size, Cohen's d, of .3 (black). Power (discussed below) is the probability of obtaining p < .05 on a one-tailed t-test comparing group means for each sample size (Based on Bishop et al., 2022.)

night for 3 months, whereas in the other group no such encouragement is given. The study is run, but when children's vocabulary results are compared, the statistical test results in a p-value of .23, much greater than the α level of .05.

The researcher is confused as she knows there is research evidence to indicate that an effect should be present. There are, of course, a number of reasons why the experiment might have turned up a null result, but anyone familiar with statistics will think of the most likely explanation: unfortunately, she has failed to take into account the fact that the effect is fairly small, and to show it convincingly she would need a much larger sample size then 10 families per group.

13.6   Statistical power and β

Statistical power is the probability that a study will show a significant difference on a statistical test when there is a true effect. Statisticians use the term β to

TABLE 13.1
Power for 2-tailed independent t-test, alpha = .05

N per group	Effect size (d)							
	0.1	0.2	0.3	0.4	0.5	0.6	0.8	1
10	0.055	0.071	0.097	0.135	0.185	0.246	0.395	0.562
20	0.061	0.095	0.152	0.234	0.338	0.456	0.693	0.869
30	0.067	0.119	0.208	0.332	0.478	0.628	0.861	0.968
40	0.073	0.143	0.263	0.424	0.598	0.755	0.942	0.993
50	0.079	0.168	0.318	0.508	0.697	0.844	0.977	0.999
80	0.096	0.242	0.471	0.710	0.882	0.965	0.999	1.000
100	0.108	0.291	0.560	0.804	0.940	0.988	1.000	1.000
150	0.139	0.408	0.736	0.932	0.991	0.999	1.000	1.000
200	0.169	0.514	0.849	0.979	0.999	1.000	1.000	1.000

refer to the proportion of nonsignificant results that are false negatives (type II error); power is 1-β, expressed either as a proportion or a percentage.

In practice, it can get confusing to think about Greek symbols (especially since β has a completely different statistical meaning in the context of regression!), but the main point to grasp is that if I say my study has 40% power, that means that, if there were a true effect of intervention, and I were to run the study 10 times, on only four occasions would I obtain a statistically significant difference.

Power depends several factors including:

- Sample size,

- True effect size in the population,

- Criterion for statistical significance, also known as the Type 1 error rate (α).

As can be seen from Table 13.1, in general, the larger the sample size, the higher the power, and the greater the effect size, the higher the power. This is a sobering table for fields where it is not uncommon to have sample sizes of 20 or less per group, especially as we know that few effective interventions have effect sizes greater than .4.

We can also show how power is affected by changing the α level – this affects how large a difference we need to see between groups before we reject the null hypothesis. When α is more extreme, we will make fewer false positive errors (see Chapter 14), but we will make more false negatives.

Figure 13.3 illustrates this in a simple example using a z-test, which simply assesses how likely it is that a sample came from a population with a given

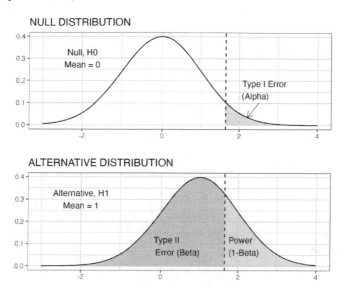

FIGURE 13.3
Z test: statistical power, N=1.

mean. This is not a realistic example, but it is used to give more insight into power. Suppose we have some background information with sets of reading scores from children who did and did not have the intervention. We now have a single child's score on a reading test, and we want to use this information to infer how likely it is that they had the reading intervention. Figure 13.3 below shows the distribution of scores for children who had no intervention in the top: this is a density plot, showing how common different scores are (proportion of the population is on the y-axis), for each specific score on the x-axis. The shape follows a normal distribution: most scores are in the middle, with higher or lower scores being less common, and the mean is zero and standard deviation is one. The null hypothesis is that the child's score comes from this distribution.

The lower figure shows the distribution for children who had the reading intervention. The intervention had a large effect (Cohen's d of one), and so the whole distribution is shifted over to the right. We're going to use a one-tailed z-test, because we know that the child's score will either come from the null distribution, or from one with a higher mean. We decide to use the conventional level of α of .05. The vertical dotted line is therefore placed at a point where 5% of the upper distribution (the red area) is to the right of the line, and 95% (the white area) to the left. This point can be worked out from knowledge of the normal distribution and corresponds to a score on the x-axis of 1.65.

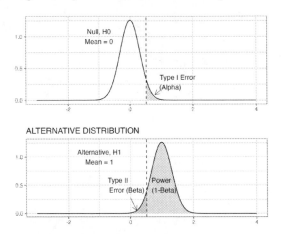

FIGURE 13.4
Z test: statistical power, N=10.

Our rule is that if the child's score is greater than 1.65, we reject the null hypothesis. If it is less than 1.65, we can't reject the null hypothesis. This does not mean that the child definitely came from the non-intervention group – just that the evidence is not sufficient to rule that out. Regardless of the effect size or the sample size, if α level is set to .05, we wrongly reject the null hypothesis on only 1 in 20 experiments.

But now look down to the lower distribution. False negatives are represented by the dark grey area to the left of the dotted line: cases where we fail to reject the null hypothesis, but the score really came from the group with intervention. The power of the study corresponds to the pale grey region, where we have a score greater than 1.65 and the data really come from the intervention condition. But notice that power is extremely low: the dark grey area is much bigger than the pale grey area. We are much more likely to wrongly retain the null hypothesis than to correctly reject it – even though we have a pretty large effect size. Note that whereas the false positive rate is kept constant by selection of α of .05, the false negative result is not.

If we wanted to make the false negative (type II error) rate much lower, we could adopt a less stringent α level; e.g., we could move the vertical dotted line to the point where the x-axis was zero, so the pale grey area becomes much bigger than the dark grey area. But if we do that, we then would increase the type I (false positive) error rate to 50%!

Figure 13.4 presents the same one-tailed z test, but here the sample size has increased to 10. A key point is that the density plot here does *not* show the distribution of scores from individual children in any one sample; it is the distribution of *means* that we would see if we repeatedly took samples of a given size. So if we had a population of 10,000 children and just kept taking samples of 10 children, each of those samples would have a mean, and it is these that are plotted here. We should notice that two things have appeared to change. First, we see a greater distinction between the two distributions. Second, we see that the critical z value (vertical dashed line) has changed location. The distributions have not changed their location (the peak of each bell shaped curve is the same), but the spread of each distribution has shrunk as a result of the increased sample size, because the precision of the estimate of the mean improves with a larger sample.The shaded areas on the density plots directly relate to the concepts outlined above: power, type I, and type II errors. When the sample size increases, the standard error (SE) reduces. We notice that the type I error rate (pale grey area in top plot) is proportionally the same at 5%, but we see a change in the two remaining quantities, power and type II error rate. This is because these quantities are linked. The area under the density curve must always remain at 1, so proportionally, we can calculate the power as 1-$β$. We can visually see this in both figures by looking at the specified areas for the alternative distribution.

If you are finding this all quite confusing, don't worry. This is complicated and even those who have statistical training can find it challenging (Bishop et al., 2022). The most important points to take away from these examples are that:

- Statistical power depends on the sample size and the effect size, as well as the level of $α$;

- With small samples, studies often have low power, meaning that even if there is a real effect, there may be little chance of detecting it;

- A p-value greater than .05 does not mean the null hypothesis is true.

It is therefore important to think about power when designing a study, or you may end up concluding an intervention is ineffective, when in fact it has a real effect that your study is underpowered to detect.

Standard Error of the Mean

It can be challenging to get an intuitive understanding of power, because the computations needed to calculate it are not straightforward. A key statistic is the standard error of the mean, also known as the SEM, usually shortened to standard error (SE). This can be thought of as an index of the variability of an estimate of the mean from a sample. If you imagine taking numerous samples from a population and estimating the mean from each one, you will end up with a distribution of means; these estimates of the mean are much more variable for small than for large samples. The SE is the standard deviation of the estimates of the mean, and it is crucially dependent on the sample size. This follows from the formula for the SE, which is computed as the SD divided by the square root of N.

The test statistic, z in this instance, which is used to derive a p-value, uses the SE as a denominator, and so will also be influenced by sample size. The z score is defined as:

$$z = (M - \mu)/SE.$$

The bigger the N, the smaller the SE, the more precise the measurement, and the higher the power of the statistical test. Note that the value entered into these equations is the *square root* of N. It follows that improvement in precision from adding extra participants to a study is greatest at small sample sizes: as shown in the figure below, the SE is approximately halved in increasing the sample size from 10 to 40, whereas changes are much smaller in going from 110 to 140 participants.

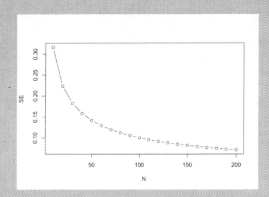

FIGURE 13.5
Plot showing how SE reduces with sample size (N) in a nonlinear fashion.

Typically, clinical trials in medicine are designed to achieve 80% statistical power and, depending on the statistical analysis strategy, will employ some method to control type I error rate (traditionally $\alpha = 0.05$). With α fixed, power depends on effect size and sample size.

So the first question is, How do we select an effect size? This is quite a thorny issue. In the past, it was common just to base anticipated effect sizes on those obtained in previous studies, but these can be overestimates because of publication bias (see Chapter 21). A more logical approach is to consider what is the smallest effect size that would be of interest: for instance, if you have a vocabulary intervention in which children start with a mean score of 20 words (SD of 10) would you be interested in an average improvement on an outcome test of half a word, or would you think the intervention would only be worthwhile if children improved on average by 4 or more words? Lakens (2021) has a useful primer on how to justify a sample size.

Once a suitable effect size is established, then it is possible to compute power for different effect sizes, to work out how big a sample would be needed to attain the desired statistical power, typically set to 80% or more.

13.7   Ways to improve statistical power

Researchers are often shocked when they first do a power analysis, to find that sample sizes that are conventionally used in their field give very low statistical power. Even more depressing, a power analysis may tell you that you would need an unfeasibly large sample in order to show an effect of interest. Researchers who start out planning a study with 20 individuals per group may be discouraged to find that they need 80 per group to do a meaningful study. This is a common predicament, but there are some ways forward:

- If the necessary sample size is too large for you to achieve, it may be worth considering forming a consortium by joining forces with other researchers. Kate Button has advocated for the adoption of **team science** in psychology, recognizing that many questions require larger samples than are typically available in any one centre (Button, 2020). A particularly innovative step has been to encourage consortia for undergraduate research projects, which, she argues, not only allows for meaningful research to be done, but also provides much better training in research methods than the conventional set-up, where each student is supposed to do an original piece of research with limited resources (Button, 2018).

- People tend to think that the only way to increase power is by getting a larger sample size, but there are other options. You may be able to improve the effect size of your outcomes by careful consideration of reliability and sensitivity of the outcome measure. Remember, effect size is the difference in means divided by the standard deviation: if you can reduce the standard deviation by minimizing random noise in your outcome measure, you will increase the effect size.

- In Chapter 14, we will consider how use of multiple outcome measures can be used to improve statistical power, provided measures are taken to avoid increasing false positives.

- It is worth consulting with a statistician about the optimal research design. An RCT with comparison of intervention and control groups is not the only possible approach. Some designs are more efficient than others for showing effects of interest: see especially Chapter 19 and Chapter 20.

13.8   Check your understanding

1. We have two studies of a spelling intervention, both using increase in number of words correct on a spelling test as the outcome measure. In both studies, two groups of 20 children were compared. In study A, the intervention group gained 10 words on average, with standard deviation of 2, and the control group gained 5 words, also with standard deviation of 2. In study B, the intervention group gained 15 words on average, with standard deviation of 5, and the control group gained 8 words with standard deviation of 5. Which study provides stronger evidence of the effectiveness of the intervention?

2. The teaching of statistics is being transformed by the internet. There are some excellent interactive websites to help you gain a more intuitive sense of some of the statistical concepts in this chapter. For starters, we'd recommend this website, which focuses on sampling from a normally distributed population: https://www.zoology.ubc.ca/~whitlock/Kingfisher/SamplingNormal.htm. You can vary the mean and standard deviation, as well as the sample size. The examples are from zoology, but the ideas apply much more broadly. This is just one of several apps in a set https://whitlockschluter.zoology.ubc.ca/stats-visualizations, that is designed to help understand statistics.

There are several online resources for calculating power for simple experimental designs. They vary in the level of statistical background that they require, but you should be able to use them to work out the required sample size for different scenarios. This site is designed to give insights into Cohen's d: https://shiny.ieis.tue.nl/d_p_power/.

Power Calculator is a site that shows how power varies with sample size and effect size: https://egap.shinyapps.io/Power_Calculator/. You should ignore the options of Clustered Design and Binary Dependent Variable for now. Also, note that when the Power Calculator asks you for the 'Treatment Effect Size' this is the *unstandardized* effect size, which is just the difference between means for Intervention and Control groups. If you specify that the standard deviation is 1, these will be equivalent to the standardized effect size, Cohen's d. Note that Power Calculator gives you the *total* number of participants. Be careful, because some power calculators give the number needed in each group.

See whether for each website you can understand the graphics, and whether you can work out the answer for the following problems, each of which is just comparing an intervention group and a control group in a two-sided t-test:

- You have an expected effect size of 0.3 and you want to work out the sample size to get power of .8 with α set to .05.

- You can recruit 50 participants per group; you expect an effect size of 0.3. What is the power to detect this effect, with α set to .05?

- What effect size would you be able to detect with 80% power and a sample size of 30 per group, with α set to .05?

More generally, it's well worth doing a web search for information if there are statistical terms or concepts you are struggling with. Sometimes the information you find is just confusing and too technical, but, as illustrated above, there are also excellent teaching resources out there.

14

False positives, p-hacking, and multiple comparisons

14.1 Learning objectives

By the end of this chapter, you will be able to:

- Explain what a type I error is and how it relates to p-hacking;
- Understand why we need to correct for multiple comparisons, and how this can be done.

14.2 Type I error

A **Type I error** is a **false positive**, which occurs when the null hypothesis is rejected but a true effect is not actually present. This would correspond to the situation where the evidence seems to indicate an intervention is effective, but in reality it has no benefit. One useful mnemonic for distinguishing Type I and Type II errors is to think of the sequence of events in the fable of the boy who cried wolf. This boy lived in a remote village, and he used to have fun running out to the farmers in the fields and shouting, "Wolf! Wolf!" and watching them all become alarmed. The farmers were making an (understandable) Type I error, of concluding there was a wolf when in fact there was none. One day, however, a wolf turned up. The boy ran into the village shouting, "Wolf! Wolf!" but nobody took any notice of him, because they were used to his tricks. Sadly, all the sheep were eaten because the villagers had made a Type II error, assuming there was no wolf when there really was one.

Figure 14.1 conveys the same information as Table 12.1 from the previous chapter, but is redrawn in terms of wolves and sheep. The type I error rate (i.e., percentage of 'significant' results seen when there is no real effect) is controlled by setting the significance criterion, $\alpha = 0.05$, at 5%, or sometimes more conservatively 1% ($\alpha = 0.01$). With $\alpha = 0.05$, on average 1 in 20

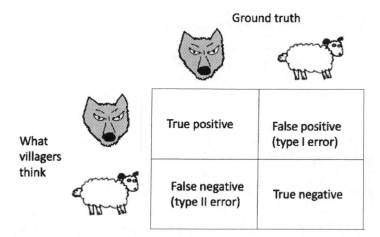

FIGURE 14.1
Type I and Type II errors.

statistically significant results will be false positives; when it is .01, then 1 in 100 will be false positives. The .05 cutoff is essentially arbitrary, but very widely adopted in the medical and social sciences literature, though debate continues as to whether a different convention should be adopted – see Benjamin et al. (2018) and Lakens et al. (2018).

14.3 Reasons for false positives

There is one obvious reason why researchers get false positive results: chance. It follows from the definition given above that if you adopt an alpha level of .05, and if the treatment really has no effect, you will wrongly conclude that your intervention is effective in 1 in 20 studies. This is why we should never put strong confidence in, or introduce major policy changes on the basis of, a single study. The probability of getting a false positive once is .05, but if you replicate your finding, then it is far more likely that it is reliable – the probability of two false positives in a row is .05 * .05 = .0025, or 1 in 400.

Unfortunately, though, false positives and non-replicable results are far more common in the literature than they would be if our scientific method was working properly. One reason, which will be covered in Chapter 21, is publication bias. Quite simply, there is a huge bias in favour of publishing papers reporting statistically significant results, with null results being quietly forgotten.

A related reason is a selective focus on positive findings *within* a study. Consider a study where the researcher measures children's skills on five different measures, comprehension, expression, mathematics, reading, and motor skills, but only one of them, comprehension, shows a statistically significant improvement (p < .05) after intervention that involves general "learning stimulation". It may seem reasonable to delete the other measures from the write-up, because they are uninteresting. Alternatively, the researcher may report all the results, but argue that there is a good reason why the intervention worked for this specific measure. Unfortunately, this would be misleading, because the statistical test needed for a study with five outcome measures is different from the one needed for a single measure. Failure to understand this point is widespread – insofar as it is recognized as a problem, it is thought of as a relatively minor issue. Let's look at this example in more detail to illustrate why it can be serious.

We assume in the example above that the researcher would have been equally likely to single out any one of the five measures, provided it gave p < .05, regardless of which one it was; with hindsight, it's usually possible to tell a good story about why the intervention was specifically effective with that task. If that is so, then interpreting a p-value for each individual measure is inappropriate, because the implicit hypothesis that is being tested is "Do *any* of these measures improve after intervention?" The probability of a false positive for *any* specific measure is 1 in 20, but the probability that *at least* one measure gives a false positive result is higher. We can work it out as follows. Let's start by assuming that in reality the intervention has no effect:

- With α set to .05, the probability that any one measure gives a *nonsignificant* result = .95.

- The probability that *all five measures* give a *nonsignificant* result is found by multiplying the probabilities for the five tasks: .95 * .95 * .95 * .95 * .95 = .77.

- So it follows that the probability that *at least one* measure gives a significant result (p-value < .05) is 1-.77 = .23.

In other words, with five measures to consider, the probability that *at least one of them* will give us p < .05 is not 1 in 20 but 1 in 4. The more measures there are, the worse it gets. We will discuss solutions to this issue below (see Multiple testing).

Psychologists have developed a term to talk about the increase in false positives that arises when people pick out results on the basis that they have a p-value of .05, regardless of the context – *p-hacking* (Simonsohn et al., 2014). Bishop & Thompson (2016) coined the term *ghost variable* to refer to a variable that was measured but was then not reported because it did not give significant results – making it impossible to tell that p-hacking had occurred. Another term, *HARKing*, or "hypothesising after the results are known" (Kerr, 1998) is used to describe the tendency to rewrite not just the Discussion but also the

Introduction of a paper to fit the result that has been obtained. Sadly, many researchers don't realize that these behaviours can dramatically increase the rate of false positives in the literature. Furthermore, they may be encouraged by senior figures to adopt exactly these practices: a notorious example is that of Bem (2004). Perhaps the most common error is to regard a p-value as a kind of inherent property of a result that reflects its importance regardless of context. In fact, context is absolutely key: a single p-value below .05 has a very different meaning in a study where you only had one outcome measure than in a study where you tested several measures in case *any* of them gave an interesting result.

A related point is that you should *never generate and test a hypothesis using the same data*. After you have run your study, you may be enthusiastic about doing further research with a specific focus on the comprehension outcome measure. That's fine, and in a new study with specific predictions about comprehension you could adopt α of .05 without any corrections for multiple testing. Problems arise when you subtly change your hypothesis *after seeing the data* from "Do *any* of these N measures show interesting results?" to "Does comprehension improve after intervention?", and apply statistics as if the other measures had not been considered.

In clinical trials research, potential for p-hacking is in theory limited by a requirement for registration of trial protocols, which usually entails that a primary outcome measure is identified before the study is started (see Chapter 22). This has not yet become standard for behavioural interventions, and indeed, clinical trial researchers often deviate from the protocol after seeing the results (Goldacre et al., 2019). It is important to be aware of the potential for a high rate of false positives when multiple outcomes are included.

Does this mean that only a single outcome can be included? The answer is no: It might be the case that the researcher requires multiple outcomes to determine the effectiveness of an intervention; for example, a quantity of interest might not be able to be measured directly, so several proxy measures are recorded to provide a composite outcome measure. But in this case, it is important to plan in advance how to conduct the analysis to avoid an increased rate of false positives.

Selection of one measure from among many is just one form of p-hacking. Another common practice has been referred to as the "garden of forking paths": the practice of trying many different analyses, including making subgroups, changing how variables are categorized, excluding certain participants post hoc, or applying different statistical tests, in the quest for something significant. This has all the problems noted above in the case of selecting from multiple measures, except it is even harder to make adjustments to the analysis to take it into account because it is often unclear exactly how many different analyses could potentially have been run. With enough analyses it is almost always possible to find something that achieves the magic cutoff of p < .05.

This animation: https://figshare.com/articles/figure/The_Garden_of_ Forking_Paths/2100379 tracks how the probability of a "significant" p-value below .05 increases as one does increasingly refined analyses of hypothetical data investigating a link between handedness and ADHD – with data split according to age, type of handedness test, gender, and whether the child's home is rural or urban. The probability of finding at least one significant result is tracked at the bottom of the display. For each binary split, the number of potential contrasts doubles, so at the end of the path there are 16 potential tests that could have been run, and the probability of *at least one* "significant" result in one combination of conditions is .56. The researcher may see that a p-value is below .05 and gleefully report that they have discovered an exciting association, but if they were looking for *any* combination of values out of numerous possibilities, then the p-value is misleading – in particular, it is *not* the case that there is only a 1 in 20 chance of obtaining a result this extreme.

There are several ways we can defend ourselves against a proliferation of false positives that results if we are too flexible in data analysis:

- Pre-registration of the study protocol, giving sufficient detail of measures and planned analyses to prevent flexible analysis – or at least make it clear when researchers have departed from the protocol. We cover pre-registration in more detail in Chapter 22.

- Using statistical methods to correct for multiple testing. Specific methods are discussed below. Note, however, that this is only effective if we correct for all the possible analyses that were considered.

- "Multiverse analysis": explicitly conducting all possible analyses to test how particular analytic decisions affected results. This is beyond the scope of this book; it is more commonly adopted in non-intervention research contexts (Steegen et al., 2016) when analysis of associations between variables is done with pre-existing data sets.

14.4   Adjusting statistics for multiple testing

As noted above, even when we have a well-designed and adequately powered study, if we collect multiple outcome variables, or if we are applying multiple tests to the same data, then we increase our chance of finding a false positive. Remember that if we set α to .05 and apply k tests to our data, then the probability of finding at least one false positive is given by $1 - (1 - \alpha)^k$. This is known as the **familywise error rate** (FWER).

Figure 14.2 shows the relationship between the familywise error rate and the number of tests conducted on the data. Note that the left-most point of the

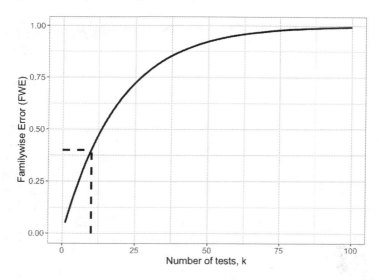

FIGURE 14.2
Plot of relationship between familywise error rate and number of statistical tests.

curved line corresponds to the case when we have just one statistical test, and the probability of a false positive is .05, exactly where we would expect. The dotted line shows the case where we performed 10 tests (k = 10), increasing our chance of obtaining a false positive to approximately 40%. Clearly, the more tests applied, the greater the increase in the chance of at least one false positive result. Although it may seem implausible that anyone would conduct 100 tests, the number of implicit tests can rapidly increase if sequential analytic decisions multiply. This is shown in the Garden of Forking Paths example, where we subdivided the analysis 4 times, to give 2^4 = 16 possible ways of dividing up the data.

There are many different ways to adjust for the multiple testing in practice. We shall discuss some relatively simple approaches that are useful in the context of intervention research, two of which can be used when evaluating published studies, as well as when planning a new study.

14.4.1 Bonferroni Correction

The Bonferroni correction is both the simplest and most popular adjustment for multiple testing. The test is described as "protecting the type I error rate;" i.e., if you want to make a false positive error only 1 in 20 studies, the Bonferroni correction specifies a new α level that is adjusted for the number of tests. The Bonferroni correction is very easy to apply: you just divide your desired false positive rate by the number of tests conducted.

For example, say we had some data and wanted to run multiple t-tests between two groups on 10 outcomes and keep the false positive rate at 1 in 20. The Bonferroni correction would adjust the α level to be $0.05/10 = 0.005$, which would indicate that the true false positive rate would be $1 - (1 - \alpha_{adjusted})^n = 1 - (1 - 0.005)^{10} = 0.049$ which approximates our desired false positive rate. So we now treat any p-values greater than .005 as non-significant. This successfully controls our type I error rate at approximately 5%.

It is not uncommon for researchers to report results both with and without Bonferroni correction – using phrases such as "the difference was significant but did not survive Bonferroni correction." This indicates misunderstanding of the statistics. The Bonferroni correction is not some kind of optional extra in an analysis that is just there to satisfy pernickety statisticians. If it is needed – as will be the case when multiple tests are conducted in the absence of clear a priori predictions – then the raw uncorrected p-values are not meaningful, and should not be reported.

The Bonferroni correction is widely used due to its simplicity but it can be over-conservative when our outcome measures are correlated. We next consider another approach, MEff, that takes correlation between measures into account.

14.4.2   MEff

The MEff statistic was developed in the field of genetics by Cheverud (2001). It is similar to the Bonferroni correction, except that instead of dividing the alpha level by the number of measures, it is divided by the *effective* number of measures, after taking into account the correlation between them. To understand how this works, consider the (unrealistic) situation when you have 5 measures but they are incredibly similar, with an average intercorrelation of .9. In that situation, they would act more like a single measure, and the effective number of variables would be close to 1. If the measures were totally uncorrelated (with average intercorrelation of 0), then the effective number of measures would be the same as the actual number of measures, 5, and so we would use the regular Bonferroni correction. In practice, the number of effective measures, MEff, can be calculated using a statistic called the eigenvalue that reflects the strength of correlation between measures. A useful tutorial for computing MEff is provided by Derringer (2018).

Bishop (2023) provided a lookup table of modified alpha levels based on MEff according to the average correlation between sets of outcome measures. Part of this table is reproduced here as Table 14.1.

This table might be useful when evaluating published studies that have not adequately corrected for multiple outcomes, provided one can estimate the degree of intercorrelation between measures.

TABLE 14.1

Adjusted alphas for multiple (N) correlated variables to achieve false positive rate of 1 in 20

	N outcomes				
Average.r	N2	N4	N6	N8	N10
0.0	0.025	0.013	0.008	0.006	0.005
0.1	0.025	0.013	0.008	0.006	0.005
0.2	0.026	0.013	0.009	0.006	0.005
0.3	0.026	0.013	0.009	0.007	0.005
0.4	0.027	0.014	0.010	0.007	0.006
0.5	0.029	0.015	0.011	0.008	0.006
0.6	0.030	0.017	0.012	0.009	0.007
0.7	0.033	0.020	0.014	0.011	0.009
0.8	0.037	0.024	0.018	0.014	0.012
0.9	0.042	0.032	0.026	0.021	0.018
1.0	0.050	0.050	0.050	0.050	0.050

14.4.3 Extracting a principal component from a set of measures

Where different outcome measures are regarded as indicators of a single underlying factor, the most effective way of balancing the risks of false positives and false negatives may be to extract a single factor from the measures, using a method such as principal component analysis. As there is just one outcome in the analysis (i.e., the score on the single extracted factor), the alpha level does not need to be modified.

Bishop (2023) used simulations to consider how statistical power was affected for a range of different effect sizes and correlations between outcomes. In most situations, power increased as more outcomes were entered into a principal component analysis (PCA), even when the correlation between outcomes was low. This makes sense, because when a single component is extracted, the resulting measure will be more reliable than the individual measures that contribute to it, and so the "noise" is reduced, making it easier to see the effect. However, unless one has the original raw data, it is not possible to apply PCA to published results. Also, the increased power of PCA is only found if the set of outcome measures can be regarded as having an underlying process in common. If there are multiple outcome measures from different domains, then MEff is preferable.

14.4.4 Improving power by use of multiple outcomes

Bishop (2023) noted that use of multiple outcome measures can be one way of improving statistical power of an intervention study. It is, however, crucial to apply appropriate statistical corrections such as Bonferroni, Principal Component Analysis, or MEff to avoid inflating the false positive rate, and it is also important to think carefully about the choice of measures: are they regarded as different indices of the same construct, or measures of different constructs that might be impacted by the intervention (see Bishop (2023) for further discussion).

14.5 Check your understanding

1. Type the phrase "did not survive Bonferroni correction" and "intervention" into a search engine to find a study that corrected for multiple corrections. Was the Bonferroni correction applied appropriately? Compare the account of the results in the Abstract and the Results section. Was the Bonferroni correction taken seriously?

2. This web-based app, https://www.shinyapps.org/apps/p-hacker/, by Schönbrodt (2016) allows you to see how easy it can be to get a "significant" result by using flexible analyses. The term DV refers to "dependent variable", i.e., an outcome measure.

- Set the True Effect slider to its default position of zero, which means that the Null Hypothesis is true – there is no true effect.

- Set the Number of DVs set to 2.

- Press "Run new experiment".

- The display of results selects the variable with the lowest p-value and highlights it in green. It also shows results for the average of the variables, as DV_all.

- Try hitting "Run new experiment" 20 times, and keep a tally of how often you get a p-value less than .05. With only two DVs, things are not so bad – on average you'll get two such values in 20 runs (though you may get more or less than this – if you are really keen, you could keep going for 100 runs to get a more stable estimate of the false positives).

- Now set the number of DVs to 10 and repeat the exercise. You will probably run out of fingers on one hand to count the number of significant p-values.

The website allows you to go even further with p-hacking, showing how easy it is to nudge results into significance by using covariates (e.g., adjusting for age, SES, etc), or by excluding outliers.

Note that if you enjoy playing with p-hacker, you can also use the app to improve your statistical intuitions about sample size and power. The app doesn't allow you to specify the case where there is only one outcome measure, which is what we really need in this case, so you have to just ignore all but one of the dependent variables (DVs). We suggest you just look at results for DV1. This time we'll assume you have a real effect. You can use the slider on True Effect to pick a value of Cohen's d, and you can also select the number of participants in each group. When you "Run new experiment" you will find that it is difficult to get a p-value below .05 if the true effect is small and/or the sample size is small, whereas with a large sample and a big effect, a significant result is highly likely.

15

Drawbacks of the two-arm RCT

15.1 Learning objectives

By the end of this chapter, you will be able to:

- Explain the difference between treatment efficacy and treatment effectiveness;
- Consider whether the RCT approach is feasible for evaluating a specific intervention.

15.2 When the RCT "gold standard" is not suitable

It can be frustrating to be told that the RCT is the "gold standard" method for evaluating interventions, only to find it is impossible to implement in many settings. Consider, for instance, the case of electropalatography, a biofeedback intervention for speech sound disorders. Leniston & Ebbels (2021) noted a number of barriers to evaluating its effectiveness through a standard RCT. The method requires expensive specialist equipment that is only available to a few speech-and-language therapists, and an individual artificial palate needs to be made for each client. The therapy is delivered over a series of sessions that can extend over several months and requires a highly trained therapist. Thus it is simply not feasible to recruit dozens, let alone hundreds, of cases for an RCT. Masking of intervention condition is also not possible; both child and therapist know who is receiving intervention. Furthermore, children who would be eligible for this intervention are likely to be very varied, both in terms of their diagnosis and associated conditions, and in terms of the specific speech sounds that they have difficulty producing. In Chapter 19 and Chapter 20, we describe alternative within-subject approaches to evaluation that can be used in this situation. Here, we elaborate further on the issues that are likely to be problematic when trying to apply RCT methods to interventions used by allied health professionals.

15.3 Inefficiency: need for unfeasibly large samples

In clinical RCTs, sample sizes typically require hundreds or even thousands of individuals, depending on analysis method and anticipated effect size, to give statistical power of 80% or more. These types of trials are usually well-funded and have a multidisciplinary team of researchers to facilitate the trial, including expert clinicians, research nurses, statisticians, data managers, field interviewers, qualitative researchers, and data entry clerks.

Historically, intervention studies in allied health professions and clinical/educational psychology have often used relatively small samples, regardless of the effect size of interest. Many of these studies were underpowered, giving a high risk of false negative results – though because of flexible data analysis and p-hacking, false positives are also common. The change to using larger well-powered studies has been somewhat slower than medical drug trial literature, in part, perhaps, because the stakes are lower: people are less likely to die or become seriously ill if the wrong conclusions are drawn from a trial. For this reason, the requirements for study design outside medical fields have been less regulated.

Suppose, however, there was a requirement that no intervention study should be approved unless the statistical power was at least 80%. If we plan to analyze outcomes using a 1-tailed t-test, then with an effect size = 0.3 – a fairly typical effect size for an effective behavioural intervention – we would need 138 cases per group. Even if we had sufficient funds to cover the staff and research costs of a study this size, for many researchers in this field, it is not possible to recruit 276 people for an intervention study.

We know that it is wasteful to run underpowered trials which have a strong likelihood of generating a type II error, so what can be done? One option, discussed in Chapter 13, is to move to a collaborative model of research where researchers pool their efforts to achieve the necessary sample size. But this is not the only way forward – and for the other reasons discussed in this chapter, it is not necessarily the optimal approach. We consider other options in chapters Chapter 19 and Chapter 20.

15.4 Transfer to real life: efficacy and effectiveness

A further issue that can hamper recruitment of an adequate sample size is the use of strict inclusion criteria. For instance, if we are studying intervention for developmental language disorder, we may wish to restrict participation to children who do not have additional conditions, such as attentional, social, or motor impairments. This serves to minimize variability among the children in the study, and avoids cases where progress is affected by other problems interfering with the intervention effect. Similar issues often crop up in medical trials, where patients with comorbidities are excluded from trials that focus on treating a particular condition. In addition, trials are usually highly controlled to ensure that randomization and experimental conditions are consistent for each participant, but this may mean that interventions are administered in a way that deviates from routine clinical practice.

This is potentially a major drawback for RCTs – in the real world, "pure" cases without comorbidities are usually rare, and quite different from the typical caseload of a practitioner. In addition, inevitably, the only patients who are entered into trials are those who are willing to be randomized, so those who have a strong preference for one intervention will not be represented. Furthermore, in practice, a lack of flexibility in administration of the intervention may deter clinicians from referring their patients to a trial if it cannot accommodate individual differences among participants.

In the clinical RCT setting, the potential mismatch between "ideal" and "real-world" (or "in-practice") settings has been much discussed, with concerns expressed that an overwhelming focus on measures taken to ensure **internal validity** of trials, (i.e., factors such as randomization and masking to avoid biases), can compromise the **external validity** of trials – i.e., ability to generalize results from highly selected patient groups to clinical reality (Rothwell, 2005). This has led to a distinction being drawn between explanatory trials (testing efficacy) or pragmatic trials (testing effectiveness). Typically in a clinical drug trial, the explanatory trial would be an earlier phase trial that tests for internal validity, and the pragmatic trial would come later, using much larger samples of individuals and testing the external validity (Patsopoulos, 2011). Singal and colleagues have summarized the notable differences between these trials (Singal et al., 2014). Table 15.1, reproduced from Singal et al. (2014), presents the key differences as summary points.

Explanatory trials (efficacy) are the most familiar trial design that is typically referred to in the RCT literature. They are designed to assess specific interventions in an idealized setting where most elements are carefully controlled and monitored. Randomization is used to ensure balance between control and intervention groups. In addition, inclusion and exclusion criteria are strictly enforced to prevent unnecessary confounding.

Pragmatic trials (effectiveness) are designed to assess interventions in a naturalistic setting which will be a far less controlled environment. Such a trial may help choose between intervention options, with a range of interventions being compared. This differs from an explanatory trial which will typically compare against a placebo condition.

TABLE 15.1

Differences between efficacy and effectiveness studies

	Efficacy/Explanatory study	Effectiveness/Pragmatic study
Question	Does the intervention work under ideal circumstances?	Does the intervention work in real-world practice?
Setting	Resource-intensive (ideal) setting	Real-world everyday clinical setting
Study population	Highly selected, homogenous population (Several exclusion criteria)	Heterogeneous population (Few to no exclusion criteria)
Providers	Highly experienced and trained	Representative usual providers
Intervention	Strictly enforced and standardized. No concurrent interventions	Applied with flexibility. Concurrent interventions and cross-over permitted

Porzsolt et al. (2015) contrasted the steps involved in an explanatory vs. pragmatic trial, noting that with a pragmatic trial, randomization is not required: eligible participants are allocated to treatment options according to individual preferences. Such a study will not control for the kind of selection biases discussed in Chapter 7 and is correspondingly regarded as a descriptive rather than explanatory trial. Stratification (comparing groups with comparable baseline scores) rather than randomization is used when analyzing results.

When designing a trial of a behavioural intervention in allied health or education, the distinction between efficacy and effectiveness is highly pertinent, and researchers need to ask themselves whether the goal is to find out what the intervention can achieve under tightly-controlled, optimal circumstances, or whether it is to test its effectiveness in a more naturalistic setting. If the decision is made to focus on "pure" cases, this may allow a more tightly controlled efficacy trial, but then questions may be raised about the generalizability of results to the broader population.

Related to this point, it is worth noting that a major reason for rejection of the term "specific language impairment" to refer to children's developmental language problems was that it encouraged a focus on children with highly selective impairments in a single domain of development, when the clinical reality is that most children with language difficulties have additional problems that may variously affect attention, social interaction, academic attainment, and/or motor skills (Bishop et al., 2017). A trial that excluded all but "pure" cases of language impairment would not only be hard to recruit for; it would also lead to results that might be difficult to generalize to the kind of child that typically featured on a clinical caseload. In a similar vein, it is important on the one hand to ensure that an intervention is delivered in an optimal fashion, as its originators intended. But on the other hand, if results are going to be generalized to the broader community, we need to show that the intervention works even in non-ideal circumstances. For instance, where intervention is administered in schools, it may be unreasonable to expect teachers to adhere strictly to a rigid timetable, because the intervention has to be integrated to fit in around other activities.

15.5   Heterogeneity and personalized intervention

In an RCT we assess the average effect of the intervention observed in a study sample. Generally, the traditional RCT is not able to provide information on treatment effectiveness for specific individuals. Indeed, if the RCT has been randomized correctly, then the identification of individuals in intervention and control groups is not possible without breaking the blinding. Hence, we can only discuss results in terms of average difference or effect size at the group level. Yet we know that people can vary in their response to intervention, and for clinical purposes, we want to know whether this intervention is likely to work for this person.

To illustrate this point, consider Figure 15.1, which shows data from a public dataset on body weight in chicks given either a regular diet (Control) or a new diet (Intervention). Average weight over time of intervention and control groups is shown as solid black and dotted lines respectively. Individual chicks are plotted as separate lines (with circles for controls and triangles for intervention), showing the variation in both groups around those averages.

This is rather far-removed from interventions used with humans, but it makes a useful point. The chicks are very homogeneous in terms of their genetic background and the environmental conditions; nevertheless, we see quite wide variation in outcomes from chick to chick. Measurement error (poor test reliability, see Chapter 3) is unlikely to explain much variation: measures of weight won't be perfectly reliable, but the differences between chicks are pretty stable across the different measurement times. So, it is clear that the birds vary in their response to the diet. Indeed, the chick with lowest weight at the final time point is from the intervention group, and some control chicks do better than the average chick in the intervention group.

In clinical trials research, there is a growing shift in focus from average results to individual response to treatment – often referred to as **precision medicine**. This has been conspicuously successful in the field of cancer treatments (Garralda et al., 2019), where knowledge of a patient's genetics has made it possible to tailor drug treatments to the individual. Two developments have followed:

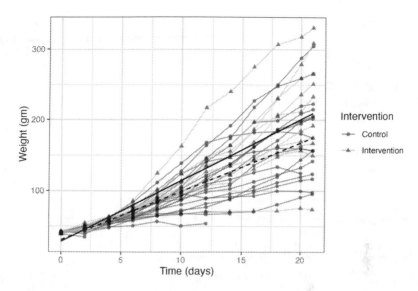

FIGURE 15.1
Data on body weight in chicks given a new diet (Intervention) or regular diet (Control). Average linear regression lines of intervention and controls group line is shown as solid black and dotted lines respectively.

(a) new methods of trial design, that are less rigid in their approach (see, e.g., Chapter 17); (b) an increased focus on biomarkers – individual differences in people's biological make-up that may be predictive of response to treatment.

It should be noted, however, that some people are sceptical about the value of biomarkers in contexts beyond oncology: precision medicine works best when there is a good understanding of the mechanism of a disease, and this is often lacking. Furthermore, if biomarkers are simply added to an analysis in the hope of finding something that predicts response to intervention, then there is a strong risk of false positive findings, unless stringent measures are taken to avoid such bias (see Chapter 14). Simply dividing people into "treatment responders" and "nonresponders" and looking for differences between them leads to many statistical traps for the unwary (Senn, 2018). Finally, the emphasis on biomarkers ignores social and environmental factors that may be important in intervention response for many diseases.

What does any of this have to do with the behavioural interventions used by allied health professionals? We should start by saying that we are dubious that the study of biomarkers will be of help in predicting response to intervention in individuals with speech and language disorders. There is a substantial research literature studying genetic and neurobiological bases of these conditions, but attempts to link this work to intervention have been plagued by methodological problems (Bishop, 2013c). A much stronger argument can be made, however, that there are individual differences between people that are likely to affect how they respond to intervention, and hence we need to develop interventions that are tailored to the individual's speech, language, and communication problems. Indeed, most speech and language therapists would argue that this is exactly what they aim to do. Insofar as this is the case, it follows that the traditional RCT has limited applicability to this field. This does not mean, however, that we should abandon attempts to evaluate interventions and simply ignore potential sources of bias. Alternative approaches that are more compatible with a "precision" approach to speech and language therapy are covered in Chapter 20.

15.6   Check your understanding

Pick an RCT on a topic of interest to you from the literature – this can be the same as for previous exercises if you wish. Consider the question (modified from one asked by Treweek & Zwarenstein (2009)):

"How likely is it that this treatment (apparently successful in this trial or review) will achieve important benefits in my context, administered by me to my patients?"

16

Moderators and mediators of intervention effects

16.1 Learning objectives

By the end of this chapter, you will be able to:

- Explain the difference between mediators and moderators;
- Identify potential mediators and moderators for interventions of interest.

16.2 Moderators

We concluded Chapter 15 by noting the heterogeneity of intervention effects, even in an experimental context where the treated individuals (chicks) had similar genetic backgrounds and upbringing. In real-life contexts with humans, we anticipate that there may be a number of factors that have a systematic influence on the effectiveness of interventions. These are referred to as **moderators**, and if we can identify them, then this will help us identify the best way of selecting individuals for intervention, either in the clinic, or in future intervention studies.

In the context of an RCT, characteristics of the individual measured in the baseline period are potential moderators. For example, in a study of children, we might divide them according to a measure of economic status, whether or not the child is eligible for free school meals. This is our moderator, M, and in this example, it just divides the sample into two subgroups.

To show that M is a moderator of treatment, we have to demonstrate that there is an interaction between M and the treatment effect, as shown in Figure 16.1. Note that the key feature is the slope of the line between baseline and outcome measurements. The overall intervention effect is estimated from the difference in the slopes for intervention and control groups. Here, the slope

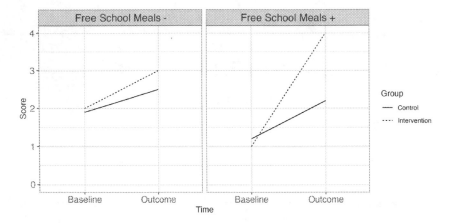

FIGURE 16.1
Interaction between intervention effect and Free School Meals status.

of the intervention effect is higher for the group with free school meals (Free School Meals +) than for the group without (Free School Meals -), suggesting that free school meals status is a moderator.

Moderators are often characteristics of the individual, such as age, sex, severity, or associated ("comorbid") conditions, and can be used to determine for whom an intervention will be most effective. Other moderators may be related to the circumstances in which the intervention is administered, e.g., whether at school or in a clinic. Morrow et al. (2022) noted that moderation analysis is particularly relevant when considering interventions used with heterogeneous populations, such as individuals who have had brain damage, where age, location, and duration of damage may have important effects on outcome. Rather than trying to correct for variation in outcome caused by these characteristics, with moderation analysis we aim to estimate which factors are important.

Although the idea of studying moderators is attractive, in practice it can be difficult to demonstrate such effects. One reason is that large sample sizes are needed to have adequate statistical power to demonstrate interaction effects (see Chapter 13), so unless moderator effects are substantial, they may be missed. Second, as noted by Kraemer et al. (2002), moderator analyses usually are exploratory rather than hypothesis-testing, and so when a moderator is discovered, it will need to be replicated in a new sample. This relates back to the arguments in Chapter 14, where we argued that you should *never generate and test a hypothesis using the same data*. Nevertheless, Kraemer et al. (2002) argue that there can be value in performing moderator analyses, as it means we can devise future interventions that build on the information gained from previous studies.

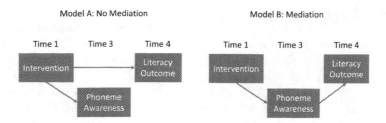

FIGURE 16.2
Models illustrating mediation vs. no mediation in reading intervention study.

16.3 Mediators

Mediators provide theoretically useful information about why and how interventions have an effect. Mediation analysis can be used as a more formal way of assessing the kinds of mechanism variables that were discussed in Chapter 4.

The principal way of distinguishing between moderators and mediators is in terms of the time of their impact. A moderator is a pre-existing variable that is independent of the allocation of individuals to intervention and control groups, but which can be shown to affect the response to intervention. In contrast, a mediator is a measured variable that changes with intervention. An example is provided by a study on reading intervention by Hulme et al. (2012). These authors contrasted two types of intervention for children with weak oral language skills. The first group (P+R) were trained in phoneme awareness, letter-sound knowledge, and guided reading of simple books. The second group (OL) had an intervention focused on oral language development, including vocabulary, grammar, and narrative skills. Outcome was measured at the end of 20 weeks of intervention, and again five months later, and it was shown that the P+R group were superior at reading. A mediation analysis was conducted to test whether improved phonemic awareness and letter-sound knowledge had led to the gains in reading. Rather than showing here the complex path diagram presented by the authors, Figure 16.2 shows two simplified models that can be compared by analysing the patterns of correlation between the measurements. In these models, arrows between variables indicate that they are significantly correlated, after taking into account all the associations between the variables. This kind of model is analyzed by using a statistical method such as path analysis, which focuses on the pattern of correlations between variables. We know from a prior analysis that there was a benefit to reading for the P+R group, and this group also showed gains in phonemic awareness, so both models include these effects. For Model A, the two effects are independent
– i.e., the intervention improves reading and improves phonemic awareness, but

the two impacts are unrelated. In this case, there is no mediation effect. For Model B, there is no direct effect of intervention on reading: instead, the effect is entirely mediated by phonemic awareness. In other words, once we have taken into account the gains shown in phonemic awareness, we have explained the gains in reading. By analysing the pattern of correlations between variables, we can see which model gives a better fit to the observed data. In the case of the Hulme et al. (2012) study, there was strong support for the mediation model. In practice, the models that are tested in mediation analysis tend to be much more complex and take into account more variables than the simplified example used here. Our goal here is to demonstrate the logic of a mediation analysis, and how it can be used to gain insights into the mechanisms underlying intervention effects.

Moderation and mediation analysis is a complicated topic, and is yet another method that can easily be misused. It is important to have a clear idea of the logic of the causal relationships between variables entered into this kind of analysis, to avoid drawing wrong conclusions. To find out more about this topic, we recommend Huntington-Klein (2023)'s book "The Effect".

16.4　　Check your understanding

Here are some characteristics of mediators and moderators. See if you can work out whether M refers to mediation or moderation.

1. In an M relationship, you can draw an arrow from intervention to M and then from M to the outcome. Mediation|moderation

2. M refers to the case when something acts upon the relationship between intevention and outcome, and changes its strength. Mediation|moderation

3. M can be thought of as a go-between for intervention and outcome: it is a variable that explains their relationship. Mediation|moderation

4. Figure 16.3 shows an intervention where both mediation and moderation are involved. Which is M1, and which is M2?

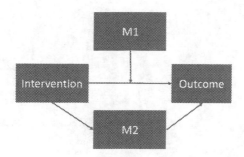

FIGURE 16.3
Path diagram with mediator and moderator.

17

Adaptive designs

17.1 Learning objectives

By the end of this chapter, you will be able to:

- Understand the characteristics of an adaptive design and recognize when it may be applicable;

- Appreciate the pros and cons of adaptive designs.

17.2 Building flexibility into trial design

Suppose we have two potential school-based reading interventions for dyslexic children, one computer game that trains sight word recognition and one phonics-based intervention. The lead researcher wishes to test both interventions but has limited funding, so can only run one trial rather than two independent trials. The researcher also has some idea that the phonics-based intervention is more likely to be effective. There is a substantial body of literature highlighting the significant effect of the phonics intervention but in younger typically developing children, whereas the computer game intervention is novel and relatively untested.

A trial design is proposed that allows both interventions to be run in parallel with the addition of a control group for reference (who receive the usual planned teaching); this is known as a three-arm trial (two interventions and a control). The researcher now has a further design decision to make, whether to use a traditional fixed design or an adaptive trial.

A traditional randomized control trial requires that data are not analyzed or "looked at" before data collection is completed. Sample size is calculated to ensure a well-powered study given certain assumptions about effect size of interest, type I error rate, and power required. The analysis is performed after data collection is complete and results reported. In contrast, in an adaptive trial, the intervention is assessed at multiple points during the recruitment

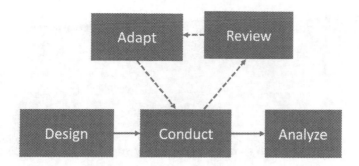

FIGURE 17.1
Flow diagram of adaptive trials.

phase of the trial without biasing the results; hence we can make certain adjustments in light of this information.

Let's return to our three-arm trial for reading interventions. The researcher has started data collection and has gathered some initial teacher feedback that the novel computer game intervention seems to be hampering children's progress. In a traditional trial there would be no other course of action but to let the trial run its course and evaluate as planned. This might lead to the computer intervention showing poor results or even having a negative impact on those children, confirming the teacher's concerns. Alternatively, if the researcher chooses an adaptive design, the teacher's observations can be assessed part way through the trial and the computer based intervention dropped from the study if interim analyses shows a negative impact of the intervention. As a consequence, the remaining individuals can change to the phonics-based intervention if the interim analyses showed positive results.

The adaptive design, as the name suggests, allows for modifications to certain aspects of the design as the trial progresses – see Figure 17.1, based on Pallmann et al. (2018). Interim analyses are pre-planned and used to assess whether modifications are required. These assessments can be used to make several pre-specified changes, if required.

Figure 17.2 (based on Thorlund et al., 2018) shows some of the commonly used adaptive trial designs. The adaptive trial with sample size reassessment has an interim step when sample size is re-evaluated based on provisional results. The trial with response adaptive randomization changes the allocation ratio to intervention and control groups depending on interim findings. The trial with adaptive enrichment changes the proportions of different participant types based on interim results. Other adaptive designs include changing the amount or frequency of intervention received, and stopping intervention early if interim analysis shows the intervention to be ineffective or if adverse effects are observed.

Conventional trial

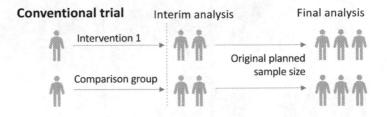

Adaptive trial with sample size reassessment (SSR)

Adaptive trial with response adaptive randomization

Adaptive trial with adaptive enrichment design

FIGURE 17.2
Commonly used adaptive designs; based on Thorlund et al., 2018.

17.3   Pros and cons of adaptive designs

Adaptive designs can be more efficient than regular trials, as they allow ineffective interventions to be dropped early. They may also be more acceptable to participants, who will appreciate that they won't be required to continue with an unpromising intervention. However, adaptive trials are not always preferable to regular trials. Before embarking on an adaptive trial, it is important to consider factors such as the time course of the study, the added complexity of planning an adaptive trial, and the possible dangers of stopping an intervention arm early on the basis of limited information. Wason et al. (2019) provides thoughtful discussion of this topic.

17.4 Check your understanding

1. A presentation by Ludlow (2013) noted the potential of adaptive designs in speech and language therapy. She noted that most trials of swallowing therapies used a similar schedule, regardless of the type of therapy: in hospital, the patient was seen 1 hour a day for 10 days, giving a total exposure of 10 hours. Alternatively, some outpatient clinics would have patients coming in for one hour per week, for 2-4 weeks. Ludlow suggested that before embarking on a large and costly RCT it would make sense to do adaptive trials to find out what was the optimal schedule.

   Suppose you had access to 30 patients who were willing to participate in a clinical trial of swallowing therapy. You decide it's worth using an adaptive design to find out optimal conditions for treatment before embarking on a larger trial. You have the potential to vary the intensity of intervention (how many sessions per day or week) and the total number of sessions. Assuming you had the resources to try any combination of these, consider how you might proceed to home in on the optimal combination.

18

Cluster Randomized Controlled Trials

18.1 Learning objectives

By the end of this chapter, you will be able to:

- Understand when a cluster RCT is preferable to a regular RCT;
- Explain the limitations of a cluster RCT.

18.2 What is a cluster RCT?

In some situations, allocation of intervention can be problematic due to pragmatic constraints; for example, for a whole-class intervention, it may be disruptive to split students within each class into intervention and control groups. A potential solution is to use **cluster trials**, where the allocation to intervention is performed at the group level. We can designate clusters that would naturally occur together, for example, school classes, patients under the care of a particular medical practice or a specific therapist, or perhaps geographic location (county, district, or healthcare provider). Figure 18.1 illustrates the logic.

18.3 Advantages of a cluster design

The first advantage of this approach is logistic. It avoids the practical difficulties of randomizing to intervention within established natural clusters, e.g., if children within a classroom are in different intervention arms, or a therapist is required to keep track of which cases are allocated to one intervention

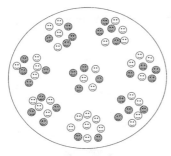
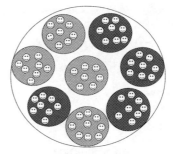

Individual randomization   Cluster randomization

FIGURE 18.1

With regular RCT (on left), individuals are sampled at random from the population for allocation to intervention (grey) or control (white). With cluster RCT (on right) natural clusters in the population (schools, hospitals, cities, etc) are sampled at random and assigned to intervention (grey) or control (black).

or another. In addition, if it is necessary to train those administering the intervention, clustering means that fewer individuals need to be trained overall, as those working in each cluster would require training in just one method.

Another important consideration is that clustering avoids "contamination" effects. Contamination in the context of controlled trials means that individuals in one arm of the trial, say the control arm, have been exposed to another arm of the trial, say intervention. Suppose in the example from the last chapter that within the same classroom we had some children who received a phonics intervention, some who received computerized intervention, and others who just had "business as usual." The children as well as the teachers might feel this is unfair, and indeed they might demand that the control children take part in the other activities. Even if there is no sense of unfairness, a teacher or therapist who is administering two interventions may find it hard to keep them separate In pharmaceutical trials, we can control contamination by masking the randomization and using placebos that are physically indistinguishable from the drug, so that neither the participant nor the researcher knows who is receiving the intervention. For behavioural interventions, this is not the case, and if participants know one another, then they may readily discover which arm of the trial they are in. The practitioner administering the intervention will also be aware of who is receiving intervention and who is not and so may treat them differently. The clustered approach minimizes the risk of contamination, as intervention clusters are less likely to have direct exposure to each other, and practitioners will consistently do the same intervention with everyone in the study.

Finally, we may note that any intervention that is administered to a whole group at a time – e.g., a modification to teacher practice that affects all children in the classroom – requires a clustered approach.

18.4   Disadvantages of a cluster design

The benefits of the clustered design are counteracted by some serious disadvantages. First of all, a clustered trial is far less efficient than a regular trial, because the unit of analysis is the cluster rather than the individual. Accordingly, much larger samples are needed to attain adequate power.

It may not be obvious why clustering should require a different analytic approach, and indeed, the literature does contain trials which have been analyzed as if they were regular trials, even though the data are clearly clustered (e.g., pupils in schools A and B receive intervention, whereas those in schools C and D are controls). The problem is that clustering introduces confounds. For instance, if we have classrooms as clusters, then all the children in one class are taught by the same teacher. They are also likely to be more similar within a cluster than between clusters on variables such as social background. This reduces variability within a cluster. Previously, we have regarded reduced variability as a good thing, as it helps pick out an intervention effect from a background of noise. But in this case, it is a bad thing, because it is confounded with the intervention effect and may make it seem larger than it is. For instance, if we just had a study where all the children in school A had intervention 1, and all of those in school B had intervention 2, then if we find children in school A did better than those in school B, this is hard to interpret. It could just be that teaching is better in school A, regardless of the intervention.

Statisticians have developed sophisticated methods of analysis for dealing with the dependency among individuals within the same cluster (Campbell & Walters, 2014). It is worth taking advice from a statistician if you are contemplating doing such a trial. For a detailed account of methods, see Gelman & Hill (2007).

18.5   Check your understanding

The Education Endowment Foundation reported on a cluster RCT of "growth mindset" intervention that was completed in 2014. The project involved 101 schools and 5018 pupils across England, assigned to either intervention or control groups.

Pupils were exposed to the mindset theory over a structured series of classes led by their own teachers. The aim of these sessions was to change the way that pupils think about their intelligence, in particular to build high expectations and resilience and to encourage specific plans and goals that emphasize growth, the development of skill, or the expansion of knowledge. The programme was delivered to pupils through eight sessions. These covered themes including the meaning of intelligence, dealing with mistakes and emotions, understanding the brain and encouraging challenge, effort and persistence, stereotypes, and inspirational people. In addition, teachers worked to make the intervention more effective by embedding the growth mindset approach in their everyday class activities, for example, by repeating the message that making mistakes is an opportunity to learn rather than a negative experience.

Teachers received one day of training that was prepared and delivered by a team of education psychologists from the University of Portsmouth. This training introduced teachers to mindset theory and evidence and provided tips on how to embed the approach in their classrooms/schools (for example, how to communicate incremental beliefs to pupils through feedback and praise). To supplement these suggested changes to everyday practice, teachers were given the materials and training to run an eight-week programme (up to 2.5 hours a week) of weekly lessons and activities with their Year 6 pupils. Specifically, teachers were given a training manual that included comprehensive lesson plans for the eight sessions and a USB stick with additional material to support their interaction with children and their parents. In addition, teachers were granted free access to online videos supplied by the Positive Edge Foundation; these videos were referenced in the lesson plans and teachers were encouraged to use the videos to explain or emphasize particular concepts such as resilience and learning from mistakes. The videos focus on three famous people (Darwin, Einstein, and Wilma Rudolph), describing their lives and how they overcame adversity, in line with the growth mindset message. They included, for example, a video about Charles Darwin's less-than-promising school performance and a medal-winning athlete who overcame a difficult start in life. Other materials provided by the project team included posters on growth mindset and pupil quizzes.

- Why was a cluster RCT method used, rather than a standard two-arm RCT?

- The results from the trial were disappointing: no evidence of any benefit was found for children who received growth mindset training. Do you think it would be worthwhile to find ways to evaluate the intervention using a non-clustered RCT? What would be the advantages/challenges of such an approach?

- The authors of the report suggested various reasons why the trial results were so disappointing. Can you think of what these reasons might be? Would there be any way of checking the plausibility of such explanations? When you have considered this point, look at the discussion of reasons for null results on pp. 42-43 of [the report], which you can find using a web search with "EEF Growth Mindset" as search terms. Did you anticipate the points raised there – and did the report consider the points you thought of? Note also how the researchers used an "implementation and process evaluation" that made it possible to evaluate different possible explanations (p. 29).

19

Cross-over designs

19.1 Learning objectives

By the end of this chapter, you will be able to:

- Understand how a cross-over RCT works;
- Specify its advantages and disadvantages compared to a regular RCT.

19.2 A within-subjects approach to the RCT

The standard two-arm RCT treats the intervention effect as a *between-subjects* factor: i.e., the intervention and control groups include different people. In the cross-over design, we treat intervention as a *within-subjects* factor, by comparing the same people in different phases of a study, during which they do or do not receive intervention. This is also known as a **repeated measures** design, because we test the same individuals in more than one condition. Because we have two groups who receive the period with and without intervention in counterbalanced order, we avoid the interpretive problems that arise with a simple pre-intervention vs. post-intervention comparison (Chapter 5).

The benefit of the crossover design is that a potentially more accurate evaluation of intervention comparison is achieved, as we compare individuals to themselves as well as to controls who are different individuals. Another benefit is that power is higher than for a regular RCT: crossover designs typically require fewer individuals, as a separate control group is not necessary. In effect, with a cross-over design, we combine two sources of information about the efficacy of intervention: comparing individuals across the two phases (*within subjects*) and comparing the two groups in each phase (*between subjects*).

In drug trials, the typical cross-over design is split into three phases: an initial phase where two groups are randomized (exactly the same as a parallel group design) to intervention and control (Sibbald & Roberts, 1998). Once the first phase has elapsed, there is a washout phase (phase 2) where no intervention is

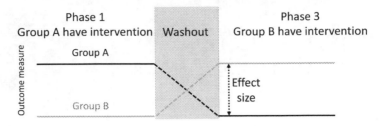

FIGURE 19.1
Schematic of a cross-over RCT.

given; this is important to allow any intervention effect to be removed before the groups are switched. Then phase 3 is started with the intervention and control conditions switched. By the end of the study, both groups have received intervention. The assessment of the intervention looks at the differences between phases 1 and 3 by group. If treatment is effective, we should see no group difference but a significant phase difference.

The suitability of a cross-over design and the anticipated results depend crucially on the nature of the intervention. A key question is whether or not the intervention is intended to have long-term effects that persist beyond the intervention phase. For most behavioural interventions, including those administered by speech-and-language therapists, educators, and allied health professionals, the goal is to bring about long-term change. Exceptions would be communication aids such as auditory feedback masking, which decreases stuttering while the masker is switched on but does not produce long-term change (Block et al., 1996). In this regard, most behavioural interventions are unlike pharmaceutical trials, which often focus on the ability of specific drugs to provide symptomatic relief. This makes results from the cross-over design difficult to interpret.

19.3   Delayed cross-over design (wait list controls)

The delayed cross-over design or wait list control design is another type of *within subject* design that is more commonly used in situations where the effect of an intervention is likely to persist. We start by taking baseline measures from both groups. The impact of intervention is measured in Group A relative to their pre-intervention score. For Group B, intervention starts at the point when Group A stops the intervention.

In many respects, this design resembles a regular RCT and has some of the same benefits, in terms of controlling for effects of practice, maturation,

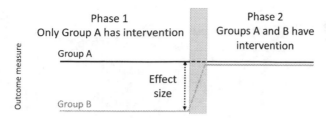

FIGURE 19.2
Schematic of a waitlist RCT.

and regression to the mean. Relative to a standard RCT, it has some advantages:

- Group B serves as a replication sample. If benefits of intervention are seen in Group A, then we should see similar effects in Group B by the end of the study.

- As well as estimating immediate effects of intervention, Group A can provide an estimate of how far there is persistence of effects by comparing their score at the end of intervention with the final, post-intervention phase.

- An adaptive approach can be adopted, so that if no benefit of intervention is seen for Group A at the end of intervention, the study can be terminated.

- This design may encourage participation by clients and those referring them, as all participants have access to a potentially beneficial intervention and serve as their own controls.

There is, however, some debate around usefulness of wait-list designs in the psychological and behavioural interventions literature (Cunningham et al., 2013; Elliott & Brown, 2002; Furukawa et al., 2014). In particular, Cunningham et al. (2013) presented results that showed this design can, potentially, artificially inflate intervention effect estimates. When a participant knows that they have to wait, this can induce a state of "resentful demoralization," which can lead to poorer outcome – a kind of **nocebo effect**. Having said that, Cunningham et al. noted that negative reactions to being on a wait list are likely to depend on the specific context. Studies showing nocebo effects have tended to focus on clinical psychology interventions, where distressed patients have been actively seeking help and may become disconsolate or irritated at having to wait.

Calder et al. (2021) used a delayed cross-over design to study the effect of an intervention designed to improve use of past tense -ed endings in children with Developmental Language Disorder. As well as comparing two groups who received intervention at different time points, they also compared different outcome measures. Here, for simplicity, we restrict consideration to the outcome measure of past tense endings – the grammatical element that the training

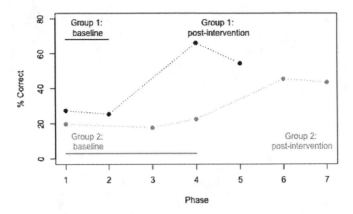

FIGURE 19.3

Mean per cent correct in delayed cross-over study by Calder et al., 2021 (data plotted from Calder et al.'s Table 2).

had focused on. (In Chapter 20 we look at how the same study incorporated aspects of single case design to obtain additional evidence that intervention benefits are specific to the skills that are trained). Results are shown in Figure 19.3.

First, we have the improvement in group 1 (black) when phases 4-5 (post-intervention) are compared with phases 1-2 (baseline). Next, look at Group 2, in grey. They provide two further pieces of evidence. First we can use them as a replication sample for Group 1, considering the change from phase 4-5 (baseline) to phase 6-7 (post-intervention): although the improvement on past-tense items is not as big as that seen in Group 1, it is evident by inspection, and again contrasts with a lack of improvement on untrained constructions. Finally, we can contrast the two groups at phases 4-5. This is more like the contrast performed in a standard RCT, a between-subjects contrast between treated vs. untreated groups at the same time point. Here, the combined evidence from between- and within-subjects comparisons provides converging support for the effectiveness of the intervention. Nevertheless, we may wonder how specific the intervention effect is. Could it be a placebo effect, whereby children's scores improve simply because they have the individual attention of an enthusiastic therapist? Maybe the children would have improved just as much if the therapist had simply spent the time reading to them or playing with them. In Chapter 20 we move on to consider outcomes for the untrained constructions, which provide evidence against that interpretation.

19.4 Check your understanding

1. Calder et al. (2021) provides a nice example of how to design a study to provide a stringent test of intervention by combining within- and between-subjects sources of evidence. It should be remembered, though, that results don't always turn out as hoped. What could you conclude if:

    - There was no difference between Groups 1 and 2 on past tense items at phase 4, with neither group showing any improvement over baseline?

    - There was no difference between Groups 1 and 2 at phase 4, with both groups showing significant improvement over baseline?

    - Group 1 improved at phase 4, but Group 2 did not improve between phase 4 and phase 6?

    - Is it realistic to consider a possible "nocebo" effect during the extended waitlist for Group 2? Is there any way to check for this?

    When planning a study of this kind, the natural tendency is to assume that everything will work out and show the intervention to be effective. It helps to design a strong study if one anticipates the possibility of messy or inconclusive results such as those described here and considers whether it is possible to design a study to avoid them.

2. Take a look at a study by Varley et al. (2016), who used a self-administered computerized intervention with stroke patients who had apraxia of speech. The intervention group did training on speech production and perception for 6 weeks, whereas the active controls were given a sham intervention that involved visuospatial processing, and were told it was designed to improve their attention and memory. After the initial intervention period, there was a 4 week break, and then the two interventions were swapped over for another 6 weeks.

    - Does this approach provide better control of internal validity than the waitlist control method used by Calder et al. (2021)?

    - Are there downsides to the use of a "sham" intervention?

20

Single case designs

20.1   Learning objectives

By the end of this chapter, you will be able to:

- Explain how different types of single case design can control for the kinds of bias discussed in previous chapters;

- Understand the limitations of single case designs for interventions that aim to have persistent effects, and how to counteract these.

20.2   Logic of single case designs

The single case design, also known as **N-of-1 trial**, or **small N design**, is a commonly used intervention design in speech and language therapy, clinical psychology, education, and neuropsychology, including aphasia therapy (Perdices & Tate, 2009). The single case design may be regarded as an extreme version of a *within-subjects* design, where two more more conditions are compared within a single person. This type of trial is sometimes dismissed as providing poor quality evidence, but a well-designed single case trial can be an efficient way to obtain an estimate of treatment efficacy in an individual. Very often, a set of single case trials is combined into a **case series** (see below). It is important to note that a single case trial is not a simple case report, but rather a study that is designed and analyzed in a way that controls as far as possible for the kind of unwanted influences on results described in Chapters 2–5.

Table 20.1 compares the logic of the standard RCT and single case designs.

The first row of Table 20.1 shows the design for a simple 2-arm RCT, where intervention is varied between participants who are assessed on the same occasion and on the same outcome. The second row shows a version of the single case design where the intervention is varied in a single subject at different time points. The third row shows the case where intervention is assessed by comparing treated vs. untreated outcomes in the same subject on the same

TABLE 20.1
How intervention vs. control is allocated in different study designs

Design	Participants	Time	Outcome
RCT: Multiple participants	Yes	-	-
Single case: Multiple time-points	-	Yes	-
Single case: Multiple outcomes	-	-	Yes

occasion – this is referred to by Krasny-Pacini & Evans (2018) as a multiple baseline design across behaviours and by Ledford et al. (2019) as an Adapted Alternating Treatment Design.

Whatever design is used, the key requirements are analogous to those of the RCT:

- To minimize unwanted variance (noise) that may mask effects of interest;

- To ensure that the effect we observe is as unbiased as possible;

- To have sufficient data to reliably detect effects of interest.

20.2.1 Minimizing unwanted variance

In the RCT, this is achieved by having a large enough sample of participants to distinguish variation associated with intervention from idiosyncratic differences between individuals, and by keeping other aspects of the trial, such as timing and outcomes, as constant as possible.

With single case trials, we do not control for variation associated with individual participant characteristics – indeed we are interested in how different people respond to intervention – but we do need to control as far as possible for other sources of variation. The **ABA** design is a popular single-case design that involves contrasting an outcome during periods of intervention (B) versus periods of no intervention (A). For example, Armson & Stuart (1998) studied the impact of frequency-altered auditory feedback on 12 people who stuttered. They contrasted a baseline period (A), a period with auditory feedback (B), and a post-intervention period (A), taking several measures of stuttering during each period. Figure 20.1 shows data from two participants during a reading condition. Although the initial amount of stuttering differs for the two individuals, in both cases there is a precipitate drop in stuttering at the 5 minute point corresponding to the onset of the masking, which is sustained for some minutes before gradually rising back towards baseline levels. The baseline period is useful for providing a set of estimates of stuttering prior to intervention, so we can see that the drop in stuttering, at least initially, is outside the range of variation that occurs spontaneously.

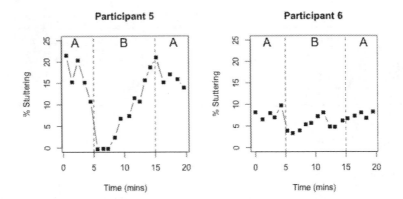

FIGURE 20.1

Outcome over time in a single case ABA design. Redrawn from digitized data from two participants from Figure 2 of Armson & Stuart (1998).

In the context of neurorehabilitation and speech-and-language therapy, there would appear to be a major drawback of the ABA design. In the course of a historical review of this approach, Mirza et al. (2017) described the "N-of-1 niche" as follows:

> *"The design is most suited to assessing interventions that act and cease to act quickly. It is particularly useful in clinical contexts in which variability in patient responses is large, when the evidence is limited, and/or when the patient differs in important ways from the people who have participated in conventional randomized controlled trials."*

While the characteristics in the second sentence fit well with speech-and-language therapy interventions, the first requirement – that the intervention should "act and cease to act quickly" is clearly inapplicable. As described in the previous chapter, with few exceptions, interventions offered by those working in education as well as speech and language therapists and those working in other allied health professions are intended to produce long-term change that persists long after the therapy has ended. Indeed, a therapy that worked only during the period of administration would not be regarded as a success. This means that ABA designs, which compare an outcome for periods with (B) and without (A) intervention, anticipating that scores will go up transiently during the intervention block, will be unsuitable. In this regard, behavioural interventions are quite different from many pharmaceutical interventions, where ABA designs are increasingly being used to compare a series of active and washout periods for a drug.

Despite this limitation, it is feasible to use an approach where we compare different time periods with and without intervention in some situations, most notably when there is good evidence that the targeted behaviour is unlikely to improve spontaneously. Inclusion of a baseline period, where behaviour is repeatedly sampled before intervention has begun, may give confidence that this is the case. An example of this **multiple baseline** approach from a study by Swain et al. (2020) is discussed below. Where the same intervention can be applied to a group of participants, then a hybrid method known as the **multiple baseline across participants** design can be used, which combines both between- and within-subjects comparisons. A study of this kind by Koutsoftas et al. (2009) is discussed at the end of this chapter.

In another kind of single case approach, the **multiple baseline across behaviours** design, it is the *outcome* measure that is varied. This approach is applicable where a single intervention has potential to target several specific behaviours or skills. This gives fields such as speech and language therapy an edge that drug trials often lack: we can change the specific outcome that is targeted by the intervention and compare it with another outcome that acts as a within-person control measure. To demonstrate effectiveness, we need to show that it is the targeted behaviour that improves, while the comparison behaviour remains unaffected.

For instance, Best et al. (2013) evaluated a cueing therapy for anomia (word-finding problems) in acquired aphasia in a case series of 16 patients, with the aim of comparing naming ability for 100 words that had been trained versus 100 untrained words. By using a large number of words, carefully selected to be of similar initial difficulty, they had sufficient data to show whether or not there was selective improvement for the trained words in individual participants.

Figure 20.2 is redrawn from data of Best et al. (2013). The black points show N items correct on the two sets of items prior to intervention, with untrained words on the X-axis and trained words on the Y-axis. The two sets were selected to be of similar difficulty, and hence they cluster around the dotted line, which shows the point where scores on both item sets are equivalent. The grey points show scores after intervention. Points that fall above the dotted line correspond to cases who did better with trained than untrained words; those below the line did better with untrained than trained words. The grey points tend to be placed vertically above the pre-test scores for each individual, indicating that there is improvement after intervention in the trained items (y-axis) but not on control items (x-axis).

Given the large number of items in each set, it is possible to do a simple comparison of proportions to see whether each person's post-intervention score is reliably higher than their pre-intervention score for each item set. For 14 of the 16 cases, there is a statistically significant increase in scores from pre-intervention to post-intervention for target items (corresponding to those with lines that extend vertically above the dotted line), whereas this is the case for only two of the cases when control items are considered (corresponding to cases which show change in the horizontal direction from pre-intervention to post-intervention).

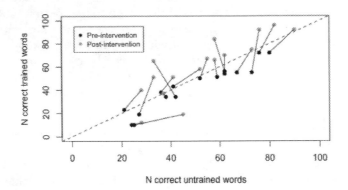

FIGURE 20.2
Outcome over time in multiple outcomes design. Reconstructed data from 16
participants, Best et al. (2013).

20.2.2   Minimizing systematic bias

We have seen in previous chapters how the RCT has evolved to minimize
numerous sources of unwanted systematic bias. We need to be alert to similar
biases affecting results of single case trials. This is a particular concern for trial
designs where we compare different time periods that do or do not include
intervention. On the one hand, we may have the kinds of time-linked effects of
maturation, practice, or spontaneous recovery that lead to a general improvement over time, regardless of the intervention (see Chapter 4), and on the other
hand there may be specific events that affect a person's performance, such as
life events or illness, which may have a prolonged beneficial or detrimental
effect on performance.

The general assumption of this method is that if we use a sufficient number of
time intervals, time-linked biases will average out, but while this may be true
for transient environmental effects, such as noise or other distractions, it is not
the case for systematic influences that continue over time. It is important to
be aware of such limitations, and it may be worth considering combining this
kind of design with other elements that control for time-related biases more
effectively (see below).

20.2.3   The need for sufficient data

Some early single case studies in neuropsychology may have drawn over-
optimistic conclusions because they had insufficient replications of outcome
measures, assuming that the observed result was a valid indication of outcome

TABLE 20.2

N items per set needed to detect target-control item difference of a given size

N_per_Set	n1	n2	n3	n4	n5	n6
10	6	13	24	43	64	81
20	10	25	49	76	92	99
30	14	37	70	92	99	100
40	17	49	81	97	100	100
50	20	58	89	100	100	100
60	25	68	95	100	100	100
70	30	73	97	100	100	100
80	32	80	98	100	100	100
90	35	85	100	100	100	100
100	38	88	100	100	100	100
110	39	90	100	100	100	100
120	44	93	100	100	100	100
130	47	94	100	100	100	100
140	49	96	100	100	100	100
150	52	97	100	100	100	100
160	56	98	100	100	100	100
170	57	98	100	100	100	100
180	60	98	100	100	100	100
190	61	99	100	100	100	100
200	66	99	100	100	100	100

without taking into account error of measurement. For instance, if someone's score improved from 2/10 items correct prior to intervention to 5/10 correct after intervention, it can be hard to draw firm conclusions on the basis of this data alone: the change could just be part of random variability in the measure. The more measurements we have in this type of study, the more confidence we can place in results: whereas in RCTs we need sufficient participants to get a sense of how much variation there is in outcomes, in single case studies we need sufficient observations, and should never rely upon just a few instances.

In effect, we need to use the same kind of logic that we saw in Chapter 13, where we estimated statistical power of a study by checking how likely we would be to get a statistically significant result from a given sample size. Table 20.2 shows power to detect a true effect of a given size in a **multiple baseline across behaviours** design of the kind used by Best et al. (2013), where we have a set of trained vs. untrained items, each of which is scored either right or wrong. The entries in this table show power, which is the probability that a study would detect a true effect of a given size on a one-tailed test. These entries were obtained by simulating 1000 datasets with each of the different combinations of sample size and effect size.

The columns show the effect size as the raw difference in proportion of items correct for trained vs. untrained words. It is assumed that these two sets were equated for difficulty prior to intervention, and the table shows the difference in proportion correct between the two sets after intervention. So if the initial proportion correct was .3 for both trained and untrained items, but after intervention, we expect accuracy on trained items to increase to .6 and the untrained to stay at .3, then the difference between the two sets after treatment is .3, shown in the 4th column of the table. We can then read down this column to see the point at which power reaches 80% or more. This occurs at the 4th row of the table, when there are 40 items in each set. If we anticipated a smaller increase in proportion correct for trained items of .2, then we would need 80 items per set to achieve 80% power.

20.3   Examples of studies using different types of single case design

As noted above, single case designs cover a wide range of options and can vary the periods of observation or the classes of observations made for each individual.

20.3.1   Multiple baseline design: Speech and language therapy for adolescents in youth justice.

The key feature of a multiple baseline design is that onset of intervention is staggered across at least three different points in time. Potentially, this could be done by having three or more participants, each of whom was measured in a baseline and an intervention phase, with the timing of the intervention phase varied across participants. Alternatively, one can have different outcomes assessed in a single participant. Figure 20.3 from Swain et al. (2020) provides an illustration of the latter approach with a single participant, where different outcomes are targeted at different points in a series of intervention sessions. Typically, the timing of the interventions is not preplanned, but rather, they are introduced in sequence, with the second intervention only started after there is a documented effect from the first intervention, and so on (Horner & Odom, 2014).

The three panels show percentages correct on outcome probes for three skills: spelling-phonics, spelling-morphology, and vocabulary. These were targeted sequentially in different sessions, and evidence for intervention effectiveness is obtained when a selective increase in performance is shown for the period

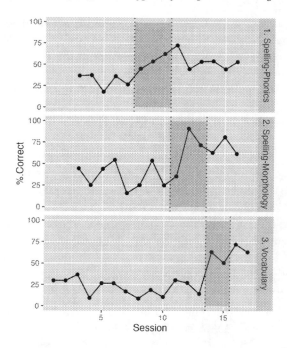

FIGURE 20.3

Data from one case from Swain et al. (2020). The shaded region shows sessions with intervention for each of the three outcomes.

during and after intervention. Note that for all three tasks, there is little or no overlap for scores during baseline and those obtained during and after intervention. The baseline data establish that although targeted behaviours vary from day to day, there is no systematic upward trend in performance until the intervention is administered. Furthermore, the fact that improvement is specific to the behaviour that is targeted in that session gives confidence that this is not just down to some general placebo effect.

In the other case studies reported by Swain et al. (2020), different behaviours were targeted, according to the specific needs of the adolescents who were studied.

20.3.2   Multiple baseline across behaviours: Effectiveness of electropalatography

We noted in the previous chapter how electropalatography, a biofeedback intervention that provides information about the position of articulators to help clients improve production of speech sounds, is ill-suited to evaluation in an RCT. It is potentially applicable to people with a wide variety of aetiologies, so the treated population is likely to be highly heterogenous, it requires expensive equipment including an individualized artificial palate, and the intervention is delivered over many one-to-one sessions. The goal of the intervention is to develop and consolidate new patterns of articulation that will persist after the intervention ends. It would not, therefore, make much sense to do a single case trial of electropalatography using an ABA design that involved comparing blocks of intervention vs. no intervention. One can, however, run a trial that tests whether there is more improvement on targeted speech sounds than on other speech sounds that are not explicitly treated.

Leniston & Ebbels (2021) applied this approach to seven adolescents with severe speech disorders, all of whom were already familiar with electropalatography. Diagnoses included verbal dyspraxia, structural abnormalities of articulators (velopharyngeal insufficiency), mosaic Turner syndrome, and right-sided hemiplegia. At the start of each school term, two sounds were identified for each case: a target sound, which would be trained, and a control sound, which was also produced incorrectly, but which was not trained. Electropalatography training was administered twice a week in 30 minute sessions. The number of terms where intervention was given ranged from 1 to 3.

An analysis of group data found no main effect of target or time, but a large interaction between these, indicating greater improvement in trained speech sounds. The design of the study made it possible to look at individual cases, which gave greater insights into variation of the impact of intervention. Figure 20.4 shows data from Leniston and Ebbels (redrawn from data kindly provided by Susan Ebbels). In the first term of intervention, there was a main effect of time for three of the participants (IDs 1, 3, and 4), but no interaction with sound type. In other words, these children improved over the course of the term, but this was seen for the untrained as well as the trained sound. By term 2, one of four children showed an interaction between time and sound type (ID4), and both children who continued training into term 3 (ID 1 and 2) showed such an interaction. Three children did not show any convincing evidence of benefit – all of these stopped intervention after one term.

As the authors noted, there is a key limitation of the study: when a significant interaction is found between time and sound type, this provides evidence that the intervention was effective. But when *both* trained and untrained sounds improve, this is ambiguous. It could mean that the intervention was effective, and its impact generalized beyond the trained sounds. But it could also mean that the intervention was ineffective, with improvement being due to other factors, such as maturation or practice on the outcome measure. Inclusion of a series of baseline measures could help establish how plausible these two possibilities were.

In sum, this method can handle the (typical) situation where intervention effects are sustained, but it is most effective if we do not expect any generalization of learning beyond the targeted behaviour or skill. Unfortunately, this is often at odds with speech and language therapy methods. For instance, in phonological

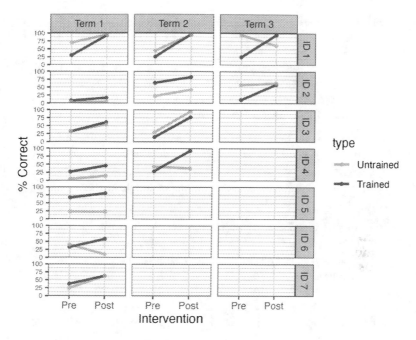

FIGURE 20.4
Individual results for targets and controls at each term.

therapy, the therapist may focus on helping a child distinguish and/or produce a specific sound pair, such as [d] vs. [g], but there are good theoretical reasons to expect that if therapy is successful, it might generalize to other sound pairs, such as [t] vs. [k], which depend on the same articulatory contrast between alveolar vs. velar place. Indeed, if we think of the child's phonology as part of a general system of contrasts, it might be expected that training on one sound pair could lead the whole system to reorganize. This is exactly what we would like to see in intervention, but it can make single case studies extremely difficult to interpret. Before designing such a study, it is worthwhile anticipating different outcomes and considering how they might be interpreted.

20.3.3   Analysis of case series data

The terms 'single case' and 'N-of-1' are misleading in implying that only one participant is trained. More commonly, studies assemble a series of N-of-1 cases. Where the same intervention is used for all cases, regular group statistics may be applied. But unlike in RCTs, heterogeneity of response is expected and needs to be documented. In fact, in a single case series, the interest is less in whether an overall intervention effect is statistically significant, as in whether the data provide evidence of individual variation in response to intervention, as this is what would justify analysis of individual cases. Formally, it is possible to test whether treatment effects vary significantly across participants by comparing a model that does or does not contain a term representing this effect, using linear mixed models, but we would recommend that researchers consult a statistician, as those methods are complex and require specific types of data. In practice, it is usually possible to judge how heterogeneous responses to intervention are by inspecting plots for individual participants.

Typically the small sample sizes in N-of-1 case series preclude any strong conclusions about the characteristics of those who do and do not show intervention effects, but results may subsequently be combined across groups, and specific hypotheses formulated about the characteristics of those who show a positive response.

An example comes from the study by Best et al. (2013) evaluating rehabilitation for anomia in acquired aphasia. As described above, researchers contrasted naming ability for words that had been trained, using a cueing approach, versus a set of untrained control words, a multiple baseline across behaviours design. In general, results were consistent with prior work in showing that improvement was largely confined to trained words. As noted above, this result allows us to draw a clear conclusion that the intervention was responsible for the improvement, but from a therapeutic perspective it was disappointing, as one might hope to see generalization to novel words.

The authors subdivided the participants according to their language profiles, and suggested that improvement on untrained words was seen in a subset of cases with a specific profile of semantic and phonological strengths. This result, however, was not striking and would need to be replicated.

20.4   Combining approaches to strengthen study design

In practice, aspects of different single-case designs can be combined – e.g., the cross-over design by Varley et al. (2016) that we described in Chapter 19 compared an intervention across two time points and two groups of participants, and also compared naming performance on three sets of items: trained words, untrained words that were phonemically similar to the trained words, and untrained words that were dissimilar to the trained words. Furthermore, baseline measures were taken in both groups to check the stability of naming responses. That study was not, however, analyzed as a single case design: rather the focus was on average outcomes without analysing individual differences. However, the inclusion of multiple outcomes and multiple time points meant that responses of individuals could also have been investigated.

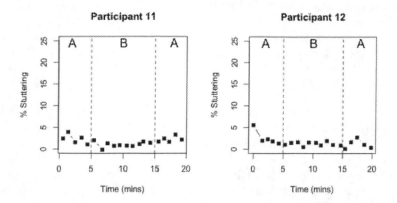

FIGURE 20.5
Outcome over time in a single case ABA design. Digitized data from two participants from Figure 2 of Armson & Stuart (1998).

20.5 Statistical approaches to single case designs

Early reports of single case studies often focused on simple visualization of results to determine intervention effects, and this is still a common practice (Perdices & Tate, 2009). This is perfectly acceptable provided that differences are very obvious, as in Figure 20.1 above. We can think back to our discussion of analysis methods for RCTs: the aim is always to ask whether the variation associated with differences in intervention is greater than the variation within the intervention condition. In Figure 20.1 there is very little overlap in the values for the intervention vs. non-intervention periods, and statistics are unnecessary. However, results can be less clearcut than this. Figure 20.5 shows data from two other participants in the study by Armson & Stuart (1998), where people may disagree about whether or not there was an intervention effect. Indeed, one criticism of the use of visual analysis in single case designs is that it is too subjective, with poor inter-rater agreement about whether effects are seen. In addition, time series data will show dependencies: autocorrelation. This can create a spurious impression of visual separation in data for different time periods (Kratochwill et al., 2014). A more quantitative approach that adopts similar logic is to measure the degree of non-overlap between distributions for datapoints associated with intervention and those from baseline or control conditions (Parker et al., 2014). This has the advantage of simplicity and relative ease of interpretation, but may be bedevilled by temporal trends in the data and have relatively low statistical power unless there are large numbers of observations.

Unfortunately, rather than a well-worked-out set of recommendations for statistical analysis of single case trials, there is a plethora of methods in use, which can be challenging, or even overwhelming, for anyone starting out in this field to navigate (Kratochwill & Levin, 2014). Furthermore, most of the focus has been on ABA and related designs, with limited advice on how to deal with designs that use comparisons between treated and untreated outcomes.

Our view is that single-case designs have considerable potential. There has been much argument about how one should analyze single case study data; multilevel models have been proposed as a useful way of answering a number of questions with a single analysis – how large the treatment effect is for individual cases, how far the effect varies across cases, and how large the average effect is. However, caution has been urged, because, as Rindskopf & Ferron (2014) noted, these more complex models make far more assumptions about the data than simpler models, and results may be misleading if they are not met. We suggest that the best way to find the optimal analysis method may be to simulate data from a single-case study design, so that one can then compare the power and efficiency of different analytic approaches, and also their robustness to aspects of the data such as departures from normality. Simulation of such data is complicated by the fact that repeated observations from a single person will show autocorrelation, but this property can be incorporated in a simulation. A start has been made on this approach: see this website: https://www.jepusto.com/easily-simulate-thousands-of-single-case-designs/ by James Pustejovsky (2018). The fact that single-case studies typically make raw data available means there is a wealth of examples that could be tested in simulations.

20.6   Overview of considerations for single case designs

In most of the examples used here, the single case design could be embedded in natural therapy sessions, include heterogeneous participants and be adapted to fit into regular clinical practice. This makes the method attractive to clinicians, but it should be noted that while incorporating evaluation into clinical activities is highly desirable, it often creates difficulties for controlling aspects of internal validity. For instance, in the study by Swain et al. (2020), the researchers noted an element of unpredictability about data collection, because the young offenders that they worked with might either be unavailable or unwilling to take part in intervention sessions on a given day. In the Leniston & Ebbels (2021) study, the target and control probes were not always well-matched at baseline, and for some children, the amount of available data was too small to give a powerful test of the intervention effect. Our view is that it is far better to aim to evaluate interventions than not to do so, provided limitations of

TABLE 20.3
How single case designs deal with threats to study validity

Biases	Remedies
Spontaneous improvement	Control outcome and/or baseline time period
Practice effects	Control outcome and/or baseline time period
Regression to the mean	Control outcome and/or baseline time period
Noisy data (1)	Baseline period to establish variability
Noisy data (2)	Outcomes with low measurement error
Selection bias	NA
Placebo effects	-
Experimenter bias (1)	Record sessions
Experimenter bias (2)	Strictly specified protocol
Biased drop-outs	Report on the number and characteristics of drop-outs
Low power	A priori power analysis; need for many observations
False positives due to p-hacking	Registration of trial protocol

particular designs are understood and discussed. Table 20.3 can be used as a checklist against which to assess characteristics of a given study, to evaluate how far internal validity has been controlled.

We are not aware of specific evidence on this point, but it seems likely that the field of single case studies, just like other fields, suffers from problems of publication bias (Chapter 21), whereby results are reported when an intervention is successful but not when it fails. If studies are adequately powered – and they should be designed so that they are – then all results should be reported, including those which may be ambiguous or negative, so that we can learn from what doesn't work, as well as from what does.

A final point, which cannot be stressed enough, is that when evaluating a given intervention, a single study is never enough. Practical constraints usually make it impossible to devise the perfect study that gives entirely unambiguous results: rather we should aim for our studies to reduce the uncertainty in our understanding of the effectiveness of intervention, with each study building on those that have gone before. With single case studies, it is common to report the raw data in the paper, in either numeric or graphical form, and this is particularly useful in allowing other researchers to combine results across studies to form stronger conclusions (see Chapter 23).

20.7   Check your understanding

1. Koutsoftas et al. (2009) conducted a study of effectiveness of phonemic awareness intervention with a group of children who showed poor sound awareness after receiving high quality whole-classroom teaching focused on this skill. Intervention sessions were administered by speech-language pathologists or experienced teachers to 13 groups of 2-4 children twice per week for a baseline and post-intervention period, and once per week during the 6 week intervention. Active intervention was preceded by a baseline period – one week (with two outcome measurement points) for seven groups of children, and two weeks (4 measurement points) for the other six groups. Outcome probes involved identifying the initial sound from a set of three words in each session. The researchers reported effect sizes for individual children that were calculated by comparing scores on the probes in the post-intervention period with those in the baseline period, showing that most children showed significant gains on the outcome measure. Group results on the outcome measure (redrawn from Table 2 of the paper) are shown in Figure 20.6.

Consider the following questions about this study.

   a. What kind of design is this?

   b. How well does this design guard against the biases shown in the first 5 rows of Table 20.3?

   c. Could the fact that intervention was delivered in small groups affect study validity? (Clue: see Chapter 18).

   d. If you were designing a study to follow up on this result, what changes might you make to the study design?

   e. What would be the logistic challenges in implementing these changes?

2. The SCRIBE guidelines have been developed to improve reporting of single case studies in the literature. An article by Tate, Perdices, Rosenkoetter, Shadish, et al. (2016) describing the guidelines with explanation and elaboration is available, as well as a shorter article

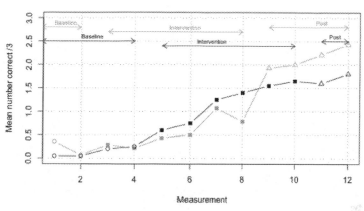

FIGURE 20.6
Group means from Koutsoftas et al., 2009. Filled points show intervention phase, unfilled show baseline or post-intervention.

summarising the guidelines (Tate, Perdices, Rosenkoetter, McDonald, et al., 2016).

Identify a single case study in the published literature in your area and check it against the guidelines to see how much of the necessary information is provided.

This kind of exercise can be more useful than just reading the guidelines, as it forces the reader to read an article carefully and consider what the guidelines mean.

3. In the previous chapter, we described a study by Calder et al. (2021), which used a cross-over design to evaluate the effect of an intervention designed to improve grammatical morphology. This study also included probes to test mastery of untrained morphological endings. The trained structure was past tense -ed; a 'generalization' probe was another verb ending, 3rd person singular -s, and a control probe was possessive -s. Before studying Figure 20.7 make a note of your predictions about what you might expect to see with these additional probes.

Once you have studied Figure 20.7, consider whether you think the inclusion of the probes has strengthened your confidence in the conclusion that the intervention is effective.

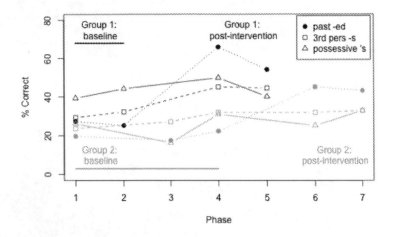

FIGURE 20.7
Mean per cent correct for all 3 probes in delayed cross-over study by Calder
et al., 2021 (data from Table 2).

21

Can you trust the published literature?

21.1 Learning objectives

By the end of this chapter, you will be able to:

- Understand how publication bias distorts the published literature, leading to intervention effects being overestimated;

- Appreciate the importance of citing negative as well as positive results when reviewing the literature on an intervention.

21.2 Publication bias

Imagine the situation of a researcher who conducts a study of the effectiveness of an intervention that she has developed to improve children's vocabulary. She has attended to all the sources of bias that feature in earlier chapters and has run a well-powered randomized controlled trial. But at the end of the day, the results are disappointing. Mean vocabulary scores of intervention and control groups are closely similar, with no evidence that the intervention was effective. So what happens next?

A common response is that the researcher decides the study was a failure and files the results away. Another possibility is that they write up the research for publication, only to find that a series of journals reject the paper, because the null results are not very interesting. There is plentiful evidence that both of these things happen frequently, but does it matter? After all, it's true that null results are uninteresting: we read journals in the hope we will hear about new, effective interventions.

In fact, failure to publish null results from well-designed studies is a massive problem for any field. This is because science is cumulative. We don't judge an intervention on a single trial: the first trial might inspire us to do further studies to get a better estimate of the effect size and generalizability of the result. But those estimates will be badly skewed if the literature is biased to contain only positive results.

This problem was recognized decades ago. Social psychologist Greenwald (1975) talked about the "Consequences of prejudice against the null hypothesis," memorably concluding that

> *"As it is functioning in at least some areas of behavioral science research, the research-publication system may be regarded as a device for systematically generating and propagating anecdotal information."*

A few years later, Rosenthal (1979) coined the term **file drawer problem** to describe the fate of studies that were deemed insufficiently exciting to publish.

In 1976, journal editor Michael Mahoney conducted a study which would raise eyebrows in current times (Mahoney, 1976). He sent 75 manuscripts out to review, but tweaked them so the results and discussion either agreed with the reviewer's presumed viewpoint or disagreed. He found that manuscripts that were identical in topic and procedure attracted favourable reviews if they found positive results, but were recommended for rejection if they found negative results. Methods that were picked apart for flaws if the findings were negative were accepted without question when there were positive results.

Michael Mahoney: A man ahead of his time

Mahoney's work has been largely forgotten, perhaps due to inaccessibility. It is in a book that is now out of print. His experience attempting to publish his study on publication bias, ironically, appears to demonstrate further evidence of such bias on behalf of an editor: *"...after completing the study on the peer review system, I submitted it to Science. After several months, I received copies of the comments of three referees. One was extremely positive, opening his review with the statement that the 'paper is certainly publishable in its present form.' The other two referees were also positive – describing it as 'bold, imaginative, and interesting' – but requesting some minor revisions in the presentation and interpretation of the data. Notwithstanding these three positive reviews, Science editor Philip H. Abelson decided to reject the manuscript! Making the minor changes mentioned by the reviewers, I resubmitted the article along with a letter to Abelson noting the positive tone of the reviews and expressing puzzlement at his decision to reject. Abelson returned a three sentence letter saying (a) the manuscript 'is obviously suitable for publication in a specialized journal', (b) 'if it were shortened it might be published as a Research Report (in Science)', and (c) that I should qualify my conclusions regarding 'the area of research and publications which are covered.' It is not clear whether this latter remark was intended to imply that the peer review system in the physical sciences is not as flawed as that in the social sciences. In any case, I shortened the article, emphasized the study's limitations, and noted the researchable possibility that different results might have been obtained with a different sample of referees or in a different discipline. My efforts were rewarded several months later when Assistant Editor John E. Ringle returned the manuscript with a five sentence rejection letter, recommending that I lengthen the article and submit it to a psychology journal."*

Figure 21.1 shows how publication bias can work out in practice. In this study, the authors searched a trial registry operated by the Food and Drug Administation (FDA) for studies on drug treatments for depression. Pharmaceutical companies are required to register intervention studies in advance, which makes it possible to detect unpublished as well as published research. As can be seen from the left-most bar (a), around half of the trials were deemed to have "negative" results, and the remainder found a positive effect of intervention. The second bar (b) shows the impact of publication bias: whereas nearly all the positive studies were published, only half of those with null results made it into print.

The next bar (c) shows yet more bias creeping in: ten negative trials switched from positive to negative, by either omitting or changing the primary study outcome. As the authors noted: *"Without access to the FDA reviews, it would*

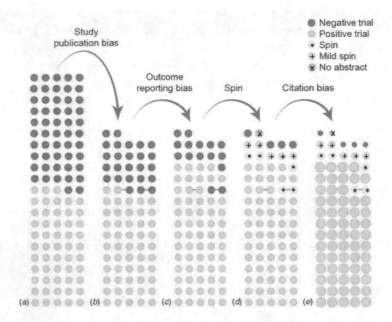

FIGURE 21.1

The cumulative impact of reporting and citation biases on the evidence base for antidepressants. Bar a displays the complete cohort of trials, while b through e show the cumulative effect of biases. Each circle indicates a trial, while the fill indicates the results or the presence of spin. Circles connected by a grey line indicate trials that were published together in a pooled publication. In e, the size of the circle indicates the relative number of citations.

not have been possible to conclude that these trials, when analyzed according to protocol, were not positive." And yet other trials reported a negative result in the body of the paper, but presented the findings with a positive spin in the Abstract, either by focusing on another result, or by ignoring statistical significance (bar (d)).

Given that the problems of publication bias have been recognized for decades, we may ask why they still persist, generating what Ferguson & Heene (2012) has termed *"A vast graveyard of undead theories."* There appear to be two driving forces. First, traditionally journals have made their money by selling content, and publishers and editors know that people are usually far more interested in positive than in negative findings. Negative findings can be newsworthy, but only only if they challenge the effectiveness of an intervention that is widely used and incurs costs. When reading the results of a positive intervention study, it is always worth asking yourself whether the study would have been accepted for publication if the results had turned out differently. For further discussion of this point see Bishop (2013b).

Another force can be conflict of interest. Sometimes the developer of an intervention has the potential to gain or lose substantial amounts of income, depending on whether a trial is positive or negative. Even if there are no financial consequences, someone who has put a lot of time and effort into developing an intervention will have a strong bias towards wanting it to work. If there is a conflict of interest, it needs to be declared: of course, we should not assume that a researcher with a conflict of interest is dishonest, but there is empirical evidence, e.g., Macnamara & Burgoyne (2023), that there is an association between conflict of interest and the reporting of positive results, which needs to be taken into account when reading the research literature.

An example of spin in evaluating speech and language therapy

A well-designed RCT (Gillam et al., 2008) included 216 school-aged children divided between four arms: (a) individualized language intervention by a speech-language pathologist; (b) computer-assisted language intervention; (c) Fast Forword-Language (FFW-L) – a commercial intervention designed to remediate a putative auditory processing deficit thought to cause language problems; (d) an academic enrichment program which did not target language skills. The last of these was intended as an active control (see Chapter 6). There was no difference in outcomes between these four groups. The natural conclusion from such a result would be that none of the language-focused interventions had succeeded in boosting language skills, because a similar amount of improvement was seen in the active control condition, which did not focus on language. Instead, the authors changed how they conceptualized the study and decided to focus on the fact that all four groups showed significant improvement from baseline to post-test. This means that the study was no longer an RCT but was rather treated as the kind of pre- vs. post-intervention design whose limitations were noted in Chapter 5. In the Discussion, the authors showed awareness of the problem of doing this: *"One important finding is that the primary measures of language and auditory processing improved significantly across groups. One possible conclusion from this pattern of results is that all four arms were equally ineffective. That is, the gains in language and auditory processing that were observed across the four arms of the study may have arisen as a result of practice effects, maturation, regression to the mean, and/or spontaneous recovery."* Yet they went on to write at length about possible reasons why the active control might have boosted language (see HARKing, Chapter 14), and the Abstract reports results in a positive fashion, noting that *"The children in all four arms improved significantly on a global language test,"* and arguing that the *improvements were of "clinical significance."* If one were to simply scan the Abstract of this paper without digging in to the Results, it would be easy to come away with the impression that a large RCT had provided support for all three language interventions.

21.3   Citation bias

In the rightmost bar of Figure 21.1, the size of the circles represents the number of citations of each paper in the literature. This illustrates another major bias that has received less attention than publication bias – citation bias. We can see that those null studies that made it through to this point, surviving with correct reporting despite publication bias, outcome reporting bias, and spin, get largely ignored, whereas studies that are reported as positive, including those which involved outcome-switching, are much more heavily cited.

Quite apart from there being bias against null results, many researchers are not particularly thorough in their citation of prior literature. In an overview of publication bias and citation bias, Leng & Leng (2020) drew attention to the neglect of prior literature in papers reporting randomized controlled trials. Based on a paper by Robinson & Goodman (2011), they noted that regardless of how many studies had been conducted, on average only two studies were cited, and concluded, *"seemingly, while scientists may see further by standing on the shoulders of those who have gone before, two shoulders are enough, however many are available"* (p. 206).

Citation bias is often unintentional, but is a consequence of the way humans think. Bishop (2020) described a particular cognitive process, confirmation bias, which makes it much easier to attend to and remember things that are aligned with our prior expectations. Confirmation bias is a natural tendency that in everyday life often serves a useful purpose in reducing our cognitive load, but which is incompatible with objective scientific thinking. We need to make explicit efforts to counteract this tendency in order to evaluate prior literature objectively.

21.4 Counteracting publication and citation biases

In Chapter 22 and Chapter 23, we discuss two approaches to counteracting biases: preregistration offers a solution to publication bias, and a systematic approach to literature review offers a (partial) solution to citation bias.

21.5 Check your understanding

1. As a group exercise, you may like to try a game of Statcheck, the materials and rules for which are available here: https://sachaeps kamp.github.io/StatcheckTheGame/(2018). This should induce a sufficiently cynical approach towards the current publication system.

2. Find a published report of an intervention that interests you. Take a look at the introduction, and list the references that are cited as background. Next, go online and search for articles on this topic that were published two or more years before the target paper. Do

you find many articles that are relevant but which were not cited? If so, did they give positive or negative results?

N.B. There are various ways you can conduct a literature search. Google Scholar is often preferred because it is free and includes a wide range of source materials, including books. Dimensions also has a free option. Scopus and Web of Science are other commercial options that your institution may subscribe to. Both of these are more selective in coverage, which can be a positive if you want some quality control over your search (e.g., restricted to peer-reviewed journals), but a negative if you want to be comprehensive. If you compare these different ways of searching the literature, you will find they can give very different results.

22

Pre-registration and Registered Reports

22.1 Learning objectives

By the end of this chapter, you will be able to:

- Understand how pre-registration of study protocols can help counteract publication bias;

- Explain how a Registered Report differs from a standard pre-registered study.

22.2 Study registration

The basic idea of study registration is that the researcher declares in advance key details of the study – effectively a full protocol that explains the research question and the methods that will be used to address it. Crucially, this should specify primary outcomes and the method of analysis, without leaving the researcher any wiggle room to tweak results to make them look more favourable. A study record in a trial registry should be public and time-stamped, and completed before any results have been collected. Since the Food and Drug Administration Modernization Act of 1997 first required FDA-regulated trials to be deposited in a registry, other areas of medicine have followed, with a general registry, https://clinicaltrials.gov/ being established in 2000. Registration of clinical trials has become widely adopted and is now required if one wants to publish clinical trial results in a reputable journal (De Angelis et al., 2004).

Study registration serves several functions, but perhaps the most important one is that it makes research studies visible, regardless of whether they obtain positive results. In Chapter 21, we showed results from a study that was able to document publication bias (De Vries et al., 2018) precisely because trials in this area were registered. Without registration, we would have no way of knowing that the unpublished trials had ever existed.

A second important function of trial registration is that it allows us to see whether researchers did what they planned to do. Of course, "The best laid schemes o' mice an' men / Gang aft a-gley," as Robert Burns put it. It may turn out to be impossible to recruit all the participants one hoped for. An outcome variable may turn out to be unsuitable for analysis. A new analytic method may come along which is much more appropriate for the study. The purpose of registration is not to put the study in a straitjacket, but rather to make it transparent when there are departures from the protocol. As noted in Chapter 14, it is not uncommon for researchers to (illegitimately) change their hypothesis on the basis of seeing the data. This practice can be misleading to both the researcher and the intended audience. It can happen that the researcher is initially disappointed by a null result but then notices that their hypothesis might be supported if a covariate is used to adjust the analysis or if a subgroup of particular individuals is analyzed instead. But if we look at the data and observe interesting patterns, then form subgroups of individuals based on these observations, we are p-hacking, and raising the likelihood that we will pursue chance findings rather than a true phenomenon (Senn, 2018). As discussed in Chapter 15, it can be entirely reasonable to suppose that some people are more responsive to the intervention than others, but there is a real risk of misinterpreting chance variation as meaningful difference if we identify subgroups only after examining the results (see Chapter 14).

Does this mean we are prohibited from exploring our data to discover unexpected findings? A common criticism of pre-registration is that it kills creativity and prevents us from making new discoveries, but this is not the case. Data exploration is an important part of scientific discovery and is to be encouraged, provided that the complete analysis timeline is presented and unregistered analyses are labelled as exploratory. An interesting-looking subgroup effect can then be followed up in a new study to see if it replicates. The problem arises when such analyses are presented as if they were part of the original plan, with results that favour an intervention effect being cherry-picked. As we saw in Chapter 14, the interpretation of statistical analyses is highly dependent on whether a specific hypothesis is tested prospectively, or whether the researcher is data-dredging – running numerous analyses in search of the elusive "significant" result; registration of the study protocol means that this distinction cannot be obscured.

In psychology, a move towards registration of studies has been largely prompted by concerns about the high rates of p-hacking and HARKing in this literature (Simmons et al., 2011) (see Chapter 14), and the focus is less on clinical trials than on basic observational or experimental studies. The term **pre-registration** has been adopted to cover this situation. For psychologists, the Open Science Framework: https://osf.io, has become the most popular repository for pre-registrations, allowing researchers to deposit a time-stamped protocol, which can be embargoed for a period of time if it is desired to keep this information private (Hardwicke & Wagenmakers, 2021). Examples of pre-registration templates are available at: https://cos.io/rr/](https://cos.io/rr/).

Does trial registration prevent outcome-switching?

A registered clinical trial protocol typically specifies a primary outcome measure, which should be used in the principal analysis of the study data. This protects against the situation where the researcher looks at numerous outcomes and picks the one that looks best – in effect p-hacking. In practice, trial registration does not always achieve its goals: (Goldacre et al., 2019) identified 76 trials published in a six-week period in one of five journals: New England Journal of Medicine, The Lancet, Journal of the American Medical Association, British Medical Journal, and Annals of Internal Medicine. These are all high-impact journals that officially endorse Consolidated Standards of Reporting Trials (CONSORT), which specify that pre-specified primary outcomes should be reported. Not only did Goldacre et al. find high rates of outcome-switching in these trial reports; they also found that some of the journals were reluctant to publish a letter that drew attention to the mismatch, with qualitative analysis demonstrating "extensive misunderstandings among journal editors about correct outcome reporting."

22.3   Registered Reports

Michael Mahoney, whose book was mentioned in Chapter 21, provided an early demonstration of publication bias with his little study of journal reviewers (Mahoney, 1976). Having found that reviewers are far too readily swayed by a paper's results, he recommended:

Manuscripts should be evaluated solely on the basis of their relevance and their methodology. Given that they ask an important question in an experimentally meaningful way, they should be published – regardless of their results (p. 105).

37 years later, Chris Chambers independently came to the same conclusion. In his editorial role at the journal Cortex, he introduced a new publishing initiative adopting this model, which was heralded by an open letter in the *Guardian* newspaper: https://www.theguardian.com/science/blog/2013/jun/05/trust-in-science-study-pre-registration. The **registered report** is a specific type of

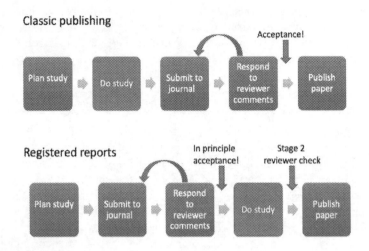

FIGURE 22.1
Comparison of stages in regular publishing model and Registered Reports.

journal article that embraces pre-registration as one element of the process; crucially, peer review occurs before data is collected.

Figure 22.1 shows how registered reports differ from the traditional publishing model. In traditional publishing, reviewers evaluate the study after it has been completed, giving ample opportunity for them to be swayed by the results. In addition, their input comes at a stage when it is too late to remedy serious flaws in the experimental design: there is a huge amount of waste that occurs when researchers put months or years into a piece of research that is then rejected because peer reviewers find serious flaws. In contrast, with registered reports, reviewers are involved at a much earlier stage. The decision whether or not to accept the article for publication is based on the introduction, methods, and analysis plan, before results are collected. At this point reviewers cannot be influenced by the results, as they have not yet been collected. The second stage review is conducted after the study has been completed, but this is much lighter touch and only checks whether the authors did as they had said they would and whether the conclusions are reasonable given the data. The "in principle" acceptance cannot be overturned by reviewers coming up with new demands at this point. This unique format turns the traditional publishing paradigm on its head. Peer review can be viewed less as "here's what you got wrong and should have done" and more like a helpful collaborator who gives feedback at a stage in the project when things can be adapted to improve the study.

The methodological quality of registered reports tends to be high because no editor or reviewer wants to commit to publish a study that is poorly conducted, underpowered, or unlikely to give a clear answer to an interesting question. Registered reports are required to specify clear hypotheses, give specification of an analysis plan to test these, justify the sample size, and document how issues such as outlier exclusion and participant selection criteria will be handled. These requirements are more rigorous than those for clinical trial registration.

The terminology in this area can be rather confusing, and it is important to distinguish between pre-registration, as described in the previous section (which in the clinical trials literature is simply referred to as "trial registration") and registered reports, which include peer review prior to data collection, with **in principle acceptance** of the paper before results are known. Another point of difference is that trial registration is always made public; that is not necessarily the case for registered reports, where the initial protocol may be deposited with the journal but not placed in the public domain. Pre-registrations outside the clinical trials domain need to be deposited on a repository with a time-stamp, but there is flexibility about when, or indeed if, they are made public.

The more stringent requirements for a registered report, versus standard pre-registration, mean that this publication model can counteract four major sources of bias in scientific publications – referred by Bishop (2019) as the four horsemen of the reproducibility apocalypse, namely:

- Publication bias. By basing reviews on introduction and methods only, it is no longer possible for knowledge of results to influence publication decisions. As Mahoney (1976) put it, it allows us to *place our trust in good questions rather than cheap answers.*

- Low power. No journal editor wants to publish an ambiguous null result that could just be the consequence of low statistical power – see Chapter 13. However, in an adequately powered intervention study, a null result is important and informative for telling us what does not work. Registered reports require authors to justify their sample size, minimizing the likelihood of type II errors.

- P-hacking. Pre-specification of the analysis plan makes transparent the distinction between pre-planned hypothesis-testing analyses and post hoc exploration of the data. Note that exploratory analyses are not precluded in a registered report, but they are reported separately, on the grounds that statistical inferences need to be handled differently in this case – see Chapter 14.

- HARKing. Because hypotheses are specified before the data are collected, it will be obvious if the researcher uses their data to generate a new hypothesis. HARKing is so common as to be normative in many fields, but it generates a high rate of false positives when a hypothesis that is only specified after seeing the data is presented as if it was the primary motivation for a study. Instead, in a registered report, authors are encouraged to present new ideas that emerge from the data in a separate section entitled "Exploratory analyses."

Registered reports are becoming increasingly popular in psychology, and are beginning to be adopted in other fields, but many journal editors have resisted adopting this format. In part this is because any novel system requires extra work, and in part because of other concerns – e.g., that this might lead to less interesting work being published in the journal. Answers to frequently asked questions about registered reports can be found on Open Science Framework: https://www.cos.io/initiatives/registered-reports. As might be gathered from this account, we are enthusiastic advocates of this approach, and have co-authored several registered reports ourselves.

22.4   Check your understanding

1. Read this preregistration of a study on Open Science Framework: https://osf.io/ecswy, and compare it with the article reporting results here: https://link.springer.com/article/10.1007/s11121-022-01455-4. Note any points where the report of the study diverges from the preregistration and consider why this might have happened. Do the changes from preregistration influence the conclusions you can draw from the study?

23

Reviewing the literature before you start

23.1 Learning objectives

By the end of this chapter, you will be able to:

- Describe the key features of a systematic review;

- Understand the main criteria used for quality appraisal in a systematic review.

23.2 Writing an Introduction

A research report typically starts with an Introduction that reviews relevant literature to provide context for the current study. It should establish what work has been done before on this topic, how much we already know, and what gaps there are in our knowledge. The primary focus of any literature review is to survey the research landscape and use this bigger picture outlook to better frame the research question of interest. The literature review may also identify methodological limitations of previous studies that can be addressed in a new study.

It may seem strange to have a chapter on the literature review at the end of this book, but there is a good reason for that. It's certainly **not** the case that we think you should only consider the Introduction only after doing the study – although that's not unknown, especially if the researcher is trying to argue for an interpretation that was not considered in advance (see HARKing, Chapter 14). Rather, it is because in order to write a good Introduction, you first need to understand a lot about experimental design and the biases that can affect research, in order to evaluate the literature properly.

Unfortunately, as noted in Chapter 21, many researchers are not objective in how they review prior literature, typically showing bias in favour of work that supports their own viewpoint. In addition, we often see distortions creeping in whereby studies are cited in a misleading fashion, omitting or de-emphasising inconvenient results. The "spin" referred to in Chapter 21 is not only applied to authors in reporting their own work: it also is seen when researchers cite the work of others.

If we accept that science should be cumulative, it is important not to disregard work that reports results that are ambiguous or inconsistent with our position, but we are often victims of our own **cognitive biases** that encourage us to disregard inconvenient facts (Bishop, 2020).

Sometimes, a failure to take note of prior work can lead to wasted research funds, as when studies are designed to address questions to which the answer is already known (Chalmers et al., 2014). In the field of speech and language therapy, where there is a shortage of published intervention research, this is unlikely to be so much of a problem. Law et al. (2003) attempted to carry out a meta-analysis to evaluate the effectiveness of intervention for children with developmental speech and language delay/disorder. Over a 25 year period, they found only 25 articles with randomized controlled designs that provided sufficient information for inclusion in their review, and the interventions used in these were quite disparate, making it difficult to combine results in a meta-analysis. Although there has been some progress in the past two decades, intervention reports in this field are still fairly rare, and few interventions could be regarded as having such a solid body of evidence in support that further studies are unnecessary. Nevertheless, it is important to take into account prior literature when planning an intervention study; yet, as Chalmers et al. (2014) noted, this is often not done.

We will start by looking at the properties of a systematic review, which is regarded as the gold standard for reviewing work in the field of health interventions.

23.3   What is a systematic review?

In clinical research, a strict definition of a systematic review is provided by the Cochrane collaboration:

> *A systematic review attempts to identify, appraise and synthesize all the empirical evidence that meets pre-specified eligibility criteria to answer a specific research question. Researchers conducting systematic reviews use explicit, systematic methods that are selected with a view aimed at*

minimizing bias, to produce more reliable findings to inform decision making.
– (Higgins et al., 2021)

The most important part of this definition is the requirement for "pre-specified eligibility criteria." In practice, many reviews that are badged as "systematic" do not conform to this requirement (Higgins et al., 2021). If a strict protocol is followed, this allows the review to be replicated or updated for a different time period. The protocol can be defined and pre-registered (see Chapter 22) and then referred to in the final published systematic review. This ensures the replicability of the review and ensures that search terms are fixed (or, in the case that search terms need to be adjusted, this can be done transparently in the write up).

Performing a systematic review according to Cochrane standards requires a group of authors in order to reduce one source of potential bias, the tendency to interpret criteria for inclusion or evaluation of studies subjectively. The level of detail required in specifying a protocol is considerable, and the time taken to complete a systematic review can be two years or more after the publication of a protocol, making this approach unfeasible for those with limited resources or time. In the field of speech and language therapy, a review of a specific intervention is unlikely to be this demanding, simply because the intervention literature is relatively small, but even so, the requirement for multiple authors may be impossible to meet, and other requirements can be onerous.

Just as with designing an intervention study, we feel that when planning a literature review, the best should not be the enemy of the good. Or as Confucius is reported to have said: *"Better a diamond with a flaw than a pebble without."* We can benefit from considering the Cochrane guidelines for a systematic review, many of which can be adopted when reviewing even a small and heterogeneous literature in order to help guard against bias. However, the guidelines, though clearly written, can be daunting because they have evolved over many years to take into account numerous considerations.

A key point is that the Cochrane advice is written to provide guidelines for those who are planning a stand-alone review that will become part of the Cochrane database. That will not be the aim of most of those reading this book; rather, the goal is to ensure that the literature review for the introduction to a study is not distorted by the usual kinds of bias that were described in Chapter 21.

Accordingly, we provide a simple summary below, adapted for this purpose. We focus on a subset of criteria used in systematic reviews, and recommend that if these can be incorporated in a study, this will make for stronger science than the more traditional style of review in which there is no control over subjective selection or interpretation of studies.

History of Cochrane

The organization now just known as Cochrane has its roots in the UK Cochrane Centre, which was founded in 1992 by Iain Chalmers and colleagues, with the aim of ensuring that medical practitioners should have access to up-to-date and objective information about results of clinical trials. Chalmers and colleagues focused on research related to management of pregnancy and childbirth, creating a database of perinatal trials which could be used as the basis for a series of systematic reviews. The value of this approach was increasingly appreciated, with the Cochrane Collaboration being formed in 1993, with 77 people from nine countries, growing over 25 years to 13,000 members and 50,000 supporters from more than 130 countries. Cochrane reviews are available online and are revised periodically to provide up-to-date summary information about a wide range of areas, principally health interventions, but also diagnostic testing, prognosis, and other topics.

23.4   Steps in the literature review

The following steps are adapted from Hemingway & Brereton (2009)'s simplified account of how to conduct a systematic review:

23.4.1   Define an appropriate question

This is often harder than anticipated. Cochrane guidelines point to the PICO mnemomic, which notes the need to specify the Population, Intervention, Comparison(s), and Outcome:

- for **Population**, you need to consider which kinds of individuals will be included in the review – e.g., age ranges, social and geographical background, diagnosis, and so on.

- For **Intervention**, one has to decide whether to adopt a broad or narrow scope. For instance, given the small amount of relevant literature, a Cochrane review by Law et al. (2003) entitled "Speech and language therapy interventions for children with primary speech and/or language disorders" took a very broad approach, including a wide range of interventions, subdivided only according to whether their focus was on speech, expressive language, or receptive language. This meant, however, that quite disparate interventions with different theoretical bases were grouped together. We anticipate that for readers of this book, the interest will be in a specific intervention, so a narrow scope is appropriate.

- **Comparison** refers to the need to specify what is the alternative to the intervention. You could specify here that only RCTs will be included in the review.

- For **Outcome**, the researcher specifies what measures of outcome will be included. These can include adverse events as well as measures of positive benefits.

Thus rather than asking "Is parent training effective in improving children's communication?", it is preferable to specify a question such as "Do RCTs show that parent training (as implemented in one or more named specific methods) leads to higher scores than for a control group on a measure of communicative responsiveness in children aged between 12 and 24 months?"

Two important recommendations by Cochrane are, first, that user engagement is important for specifying a question; this helps avoid research that will not be of practical value because its focus is too academic or narrow. Second, in considering outcomes, it is recommended that consideration be given to possible negative consequences as well as potential benefits of the intervention.

23.4.2 Specify criteria for a literature search

The goal is to identify all relevant evidence on the research question of interest, and the usual approach is to start by specifying keywords that will be entered into a literature search. Again, this might seem easy, but in practice it requires expertise, both in identifying relevant keywords and in selecting suitable databases for the search. It is worth taking advice from an academic librarian on this point.

In addition, we know that publication bias occurs, and so efforts need to be made to find relevant unpublished studies, that may be written up in preprints or theses. The search often proceeds through different stages, starting with a formal literature search, but then adding additional studies found by perusing the reference list of relevant papers. At this point, it is better to be overinclusive than underinclusive.

23.4.3 Select eligible studies following the protocol's inclusion criteria

The next step is a screening process, which typically whittles down the list of potential studies for inclusion by a substantial amount. Abstracts for the studies found in the previous step are scrutinized to see which should be retained. For example, if the protocol specified that studies had to be randomized controlled trials, then any study that did not include randomization or that lacked a

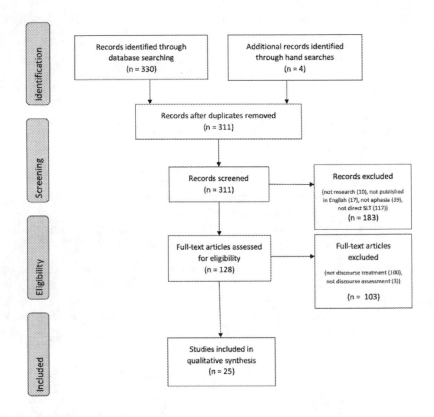

FIGURE 23.1
PRISMA flowchart from Dipper et al. (2020).

control group would be excluded at this stage. This is a key step where it is good to have more than one person doing the screening, to reduce subjective bias and to minimize the workload.

In a freestanding systematic review, it is customary to provide a flow chart documenting the numbers of papers considered at each stage. A template for a standard flow chart can be downloaded from the PRISMA website: http://prisma-statement.org/prismastatement/flowdiagram.aspx.

Figure 23.1 shows a completed flowchart for a study of intervention to improve discourse in patients with aphasia (Dipper et al., 2020); the initial search found 334 records, but on further screening, only 25 of these met eligibility criteria. The need to screen considerably more articles than the final included set is one factor that makes systematic reviews so time-consuming. It may be possible to enlist students to help with screening. For those who are familiar with the computer language R, the *metagear* package in R (Lajeunesse, 2016) is very useful for speeding up screening, by automating part of the process of extracting and evaluating abstracts for suitability.

23.4.4 Assess the quality of each study included

At this stage the methodology of each study is scrutinized to see whether it meets predefined quality criteria. Some studies may be excluded at this stage; for others, a rating is made of quality so that higher weighting can be given to those with high quality – i.e., those that have been designed to minimize the kinds of bias we have focused on in this book.

A wide range of checklists has been developed to help with quality appraisal, some of which are lengthy and complex. The questionnaires available from the Centre for Evidence-Based Management (CEBMa) https://cebma.org/resources-and-tools/what-is-critical-appraisal/ are reasonably user-friendly and include versions for different study types.

In our experience, evaluating quality of prior research is the most difficult and time-consuming aspect of a literature review, but working through a study to complete a CEBMa questionnaire can be a real eye-opener, because it often reveals problems with studies that are not at all evident when one just reads the Abstract. Researchers may be at pains to hide the weaknesses in their study, and all-too-often inconvenient information is just omitted. There may be substantial discrepancies between what is stated in the Abstract and the Results section – classic spin (see Chapter 21). Evaluating quality can be a very valuable exercise for the researcher, not just for deciding how much faith to place in study results, but also for identifying shortcomings of research that can, hopefully, be avoided in one's own study.

One can see that this stage is crucial and, even with well-defined quality criteria, there could be scope for subjective bias, so ideally two raters should evaluate each study independently so the amount of agreement can be recorded. Disagreements can usually be resolved by discussion. Even so, a sole researcher can produce a useful review by following these guidelines, provided that the process is documented transparently, with reasons for specific decisions being recorded.

23.4.5 Organize, summarize, and unbiasedly present the findings from all included studies

For a Cochrane systematic review, the analysis will have been prespecified and pre-registered. In medical trials, a commmon goal is to extract similar information from each trial on effect sizes and combine this in a meta-analysis, which provides a statistical approach to integrating findings across different studies. A full account of meta-analysis is beyond the scope of this book, but for further information, see Borenstein et al. (2009). Worked examples of meta-analysis in R are available online: https://ebmh.bmj.com/content/22/4/153.abstract.

For the typical literature review in the field of speech and language therapy, the literature is likely to be too small and/or heterogeneous for meaningful meta-analysis, and narrative review then makes more sense; i.e., the author describes the results of the different studies, noting factors that might be looked at in the future as potential influences of outcomes. Findings from the quality evaluation should be commented on here to help readers understand why some studies are given more weight than others when pulling the literature together.

23.5   Systematic review or literature review?

As we have seen, there are critical differences between a formal systematic review that follows Cochrane methods and a literature review. A well-executed systematic review is considered to be one of the most reliable sources of evidence beyond the RCT and is specific to a particular research question. It will often incorporate a meta-analytic procedure, and the end result will be a stand-alone publication.

Traditionally, the literature review that is included in the introduction to a research article is less structured and focuses on qualitative appraisal of either a general topic or specific question in the literature. Systematic reviews have tended to be seen as quite separate from more traditional literature reviews, because of the rigorous set of procedures that must be followed, but we have argued here that it is possible to use insights from systematic review methods, even if we do not intend to follow all the criteria that are required for that approach, as this will improve the rigour of the review and help avoid the kind of citation bias that pervades most fields.

When writing up a literature review, the key thing is to be aware of a natural tendency to do two things, which map rather well onto our earlier discussions of random and systematic bias. Random bias, or noise, gets into a literature review when researchers are simply unscholarly and sloppy. We are all very busy, and it is tempting to just read the Abstract of a paper rather than wade through the whole article. This is dangerous, because Abstracts often are selective in presenting the "best face" of the results, rather than the whole picture. Learning to do a good quality evaluation of a paper is an important skill for any researcher.

The second natural tendency is to show systematic bias, as discussed in Chapter 21. We are naturally inclined to look only for studies that support our pre-existing views. If we have a mixture of studies, some supportive and others not, we are likely to look far more critically at those whose results we would like to disregard. By following the steps of literature review listed above, we are helped to look at the evidence in an objective way.

23.6   Check your understanding

If possible, it is useful to do this exercise in groups of 2-4. The group should identify a paper that evaluates an intervention, and then consult this website, https://cebma.org/resources-and-tools/what-is-critical-appraisal/, to find a questionnaire that matches the methodology.

Each group member should first spend 20-30 minutes attempting to complete the questionnaire independently for the paper. Then come together and compare notes and discuss:

- Were there points of disagreement? If so, what caused these?

- How easy was it to evaluate the paper?

- Was crucial information missing? (It often is!) – Did the Abstract give an accurate indication of the Results of the paper?

24

Putting it all together

In this chapter we provide a list of questions that can be used for for evaluating published studies, and which may also be useful when planning a new study.

24.1 What is the conceptual model of the study?

A good study will have a clearly articulated conceptual model, i.e., be able to state what it is that the intervention is trying to change (the dependent variable), what is being manipulated to bring about that change (the intervention), what is the presumed mechanism of change, and what is the expected time course of change (Horner & Odom, 2014). The conceptual model is key to selecting an appropriate research design. If it is not explicitly stated, it should be possible to identify it from the study introduction and methods.

24.2 What is being compared with what in this study?

Table 20.1 in Chapter 20 provides a simple framework. There are three basic options when evaluating an intervention:

- A. Different people (typically treated or control) are compared at one time point (usually after intervention);

- B. The same person or people are compared at two or more time points (usually before and after intervention);

- C. The same person is compared on different outcome measures, only some of which have been the focus of intervention.

Option A corresponds to the classical RCT, and, if properly designed it controls for many biases; see Table 12.1 in Chapter 10. Although the focus is on comparisons after intervention, it is usual to also adjust the score for baseline levels, thereby including some elements of Option B. Option B on its own is usually a weak design because, as shown in Chapter 5, it fails to control for many biases that are related to the simple passage of time, though single case designs that use multiple baseline measures can counteract this.

Option C is sometimes used in single case designs. Our view is that it has potential for improving power and providing converging evidence for intervention effects, especially when combined with A and/or B.

24.3 How adequate are the outcome measures?

Here we return to factors considered in Chapter 3; our impression is that many studies give insufficient attention to outcome measures, which need to be reliable, valid, sensitive, and efficient; instead people tend to use measures that are available and that they are familiar with. There is a need for more systematic studies that compare the suitability of different outcome measures for intervention research, bearing in mind that measures that are useful for assessment and diagnosis may not be optimal as outcome measures.

24.4 How adequate is the sample size?

Unfortunately, many studies are simply too small to demonstrate intervention effects. Enthusiasm to show that an intervention works often propels people into doing studies that may involve considerable time, money, and resources, yet are unlikely to show an effect, even if it is really there. If you read a report of an intervention that finds a null result, it is important to consider whether the statistical power was adequate (see Chapter 13).

24.5 Who took part in the study?

As well as sample size, we need to consider whether study participants were representative of the population of interest. One cannot coerce people into taking part, and there will always be a degree of self-selection bias, but it should be possible to tell from background measures and demographic information how representative the sample is of the population of interest. In addition, the report of the study should document how many drop-outs there were, and how this was handled (see Chapter 9).

24.6   Was there adequate control for bias?

Here one wants to know whether steps were taken to avoid biases that might arise if experimenters have a strong desire for an intervention to be effective, which might influence the outcome measures or the write-up of the study. Was selection to intervention/control groups randomized? Could subjective judgments by experimenters have affected the results? Is any conflict of interest declared?

24.7   Was the study pre-registered?

As we discussed in Chapter 22, study registration is widely used in clinical medicine as a method to reduce analytic flexibility that can give misleading results (see Chapter 14). However, we know that, even when a study is pre-registered, researchers may depart from the analytic plan, and so it can be illuminating to compare the registration with the published paper.

24.8   Was the data analysis appropriate?

We have not gone into detail regarding technical aspects of data analysis, and there are plenty of useful texts that cover this ground. Even without detailed statistical knowledge, one can ask whether there is evidence of p-hacking (selectively reporting only those results that are 'significant'), and whether the presentation of the results is clear. Studies that simply report tables of regression coefficients and/or p-values are not very useful for the clinician, who will want to have a more concrete idea of how much change might be associated with an intervention, in order to judge whether it is cost-effective.

24.9   Is the data openly available?

Open data, like preregistration, doesn't guarantee high quality research, but it is a sign that the researchers are aware of the importance of open, reproducible practices, and it provides an opportunity for others to check results and/or incorporate them in a meta-analysis.

24.10   How do the results compare with others in the literature?

We have emphasized that a single study is never conclusive. One needs to combine information from a range of sources. It is surprisingly common, though, for intervention studies to be written up as if they stand alone – perhaps because many journals put such emphasis on novelty. Trustworthy work will situate results in the context of other studies and discuss potential explanations for any discrepancies in findings. When comparing studies, we need to move away from a simple binary divide between 'significant' and 'nonsignificant' findings, to consider whether effect sizes are similar, and if not, why not.

24.11   Summing up

We hope that the material in this book will give readers confidence to scrutinize the intervention literature in this way, and appreciate that it is possible to evaluate important aspects of a study without requiring advanced skills in statistics. Evaluating what works is a key skill for anyone who delivers interventions, so that the most effective approaches can be identified, and future studies can be planned to be rigorous and informative.

Comments on exercises

Chapter 1

1. *We are measuring people's weight, but the scales are incorrectly calibrated, so the true measurement is always underestimated*

   Comment: this is a systematic bias, because the defective scales lead to a bias in one direction. Note that scales that overestimated weight would also give systematic bias, whereas scales that were just erratic, and so gave very unstable readings, would be a source of random error.

2. *To get an index of social responsiveness, children's interactions with others are measured in the classroom. The researcher makes one observation per child, with different children assessed at different times of day during different activities.*

   Comment: this method is likely to create random error, because social interaction may be greater with some activities than others. Another way to think about it is that the measure will be 'noisier', i.e., more variable, than it would be if the researcher was able to keep the context of the observations more constant from child to child.

3. *In an online experiment where we measure children's keypress responses on a comprehension test, the internet connection is poor and so drops out intermittently*

   Comment: this is another context that will just increase the random error in the measures, but should not cause systematic bias towards fast or slow measures, provided the drop-out is random. On the other hand, if we were measuring children's comprehension before and after intervention, and the signal dropout was only a problem during one of the test sessions, then this could introduce systematic bias, because only one set of measures (baseline or outcome) would be affected.

4. *In an intervention study with aphasic individuals, prior to treatment, vocabulary is measured on a naming test by a therapist in the clinic,*

and at follow-up it is measured at home by a carer who gives clues to the correct answer.

Comment: This method is likely to produce systematic bias, because the conditions are more likely to lead to correct performance in the follow-up session. This could give an artificially inflated estimate of the effectiveness of intervention. In general, it is advisable to keep testing conditions as similar as possible for baseline and outcome measures. Note, though, that even if we used identical procedures at baseline and outcome, we might still have systematic bias associated with to the passage of time, due to spontaneous recovery and greater familiarity with the naming task. This is considered further in Chapter 3.

Chapter 2

2. *The two bars on the left show results obtained in an experimental study where assignment to surgery was done at random, whereas the bars on the right show results from an observational study of patients who were not included in a trial. What do these plots show, and how might it be explained?*

Comment: In Chapter 10, we consider how a randomized trial aims to ensure we are comparing like with like, by assigning people to interventions at random, but there will be people who do not get included in the trial because they are unsuitable, or because they do not wish to take part, and their outcomes may be systematically different from those who take part in the trial. In this example, patients who are at high risk may not be invited to participate in a trial of surgery because they are too frail to undergo the operation. These people feature in the observational study on the right hand side of the plot, but are filtered out of the experimental trial as they do not meet criteria for inclusion.

Chapter 3

1. *A simple way to measure children's language development is in terms of utterance length. Roger Brown's (Brown, 1973) classic work showed that in young children Mean Length of Utterance in morphemes (MLU) is a pretty good indicator of a child's language*

level; this measure counts each part of speech (morpheme) as an element, as well as each word, so the utterance "he wants juice" has 4 morphemes (he + want + s + juice). Brown's findings have stood the test of time, when much larger samples have been assessed. Is this evidence of reliability, validity, and/or sensitivity?

Comment: The plot does not give any information about reliability – we can only assess that if we have more than one MLU measurement on a set of children, so we can see how much they change from one occasion to the next. The plot does support the validity of MLU as a measure of language development: we would expect language skills to increase with age, and this plot shows that they do on this measure. MLU seems to be more sensitive as an index in younger children, where the slope of the increase with age is steeper. Indeed, it has been suggested that it is less useful in children aged 4 years and over, though this may depend on whether one is taking spontaneous language samples, or attempting to elicit more complex language – e.g., in a narrative context. To get a clearer evaluation of the psychometric properties of MLU, we would need to plot not just the changes in mean MLU with age, but also the spread of scores around the mean at each age. If there is a wide range of MLU at a given age band, this suggests the measure may be sensitive at differentiating children with language difficulties from the remainder – though this would only be the case if the measure was also shown to be reliable. If MLUs at a given age are all bunched together, then this shows that the measure does not differentiate between children, and it is unlikely to be a sensitive measure for an intervention study.

2. *Here we focus on* **reliability**, *i.e., the likelihood that you might see similar results if you did the same assessment on two different occasions. How might you expect the reliability of MLU to depend on:*

   - *Length of language sample?*

   - *Whether the child is from a clinical or typically-developing sample?*

   - *Whether the language sample comes from an interaction with a caregiver vs. an interaction with an unfamiliar person?*

Comment: In general, the more data that is used for a measure, the more reliable it is likely to be. Suppose you had just 10 utterances for computing MLU. Then one very long utterance could exert a big impact on the MLU, whereas the effect would be less if the mean were based on 100 utterances. In terms of the nature of the sample, there's no general rule, except to say that if a child from a clinical sample produces fewer utterances, then MLU will be based on fewer observations, so may be less reliable. Similarly, it's hard to predict just how the type of interlocutor will affect a child's utterances. If a child is shy in the presence of an unfamiliar person, it may be harder to get a good sample of utterances, and the MLU may change over time. On the other hand, someone who is skilled at interacting with children may be better at eliciting complex language from a child who would otherwise tend to be monosyllabic. The point here is to emphasize that psychometric properties of measures are not set in stone, but can depend on the context of measurement.

3. *Do a literature search to find out what is known about test-retest reliability of MLU. Did your answers to question 2 agree with the published evidence?*

   Comment: You will find that tracking down relevant information is quite a major task, but it can be eye-opening to do such an exercise, and it is recommended to do a search for research on outcome measures before embarking on a study that uses them.

4. *You have a three-year-old child with an MLU in morphemes of 2.0. Is there enough information in Figure 3.5 to convert this to a standard score?*

   Comment: The answer is no, because you only have information on the mean MLU at each age. To compute a standard score you would also need to know the standard deviation at each age.

Chapter 4

1. *In their analysis of the original data on the Hawthorne effect, Levitt & List (2011) found that output rose sharply on Mondays, regardless of whether artificial light was altered. Should we be concerned about possible effects of the day of the week or the time of day on intervention studies? For instance: Would it matter if all participants were given a baseline assessment on Monday and an outcome assessment on Friday? Or if those in the control group were tested in the afternoon, but those in the intervention group were tested in the morning?*

Comment: In Chapter 5, we discuss further how designs that just compare the same individuals before and after intervention are likely to be misleading, in part because time is an unavoidable confound. Time of day can be important in some kinds of biomedical studies, and with children it may be related to attentiveness. In general, it would be advisable to avoid the kind of morning vs. afternoon confound in the second example – as we shall see in Chapter 6, with a strong research design, the experimenter will be unaware of whether someone is in an intervention or control group, and so this problem should not arise. Day of the week is seldom controlled in intervention studies, and it could pose logistical problems to try to do so. Here again, a study design with a control group should take care of the problem, as any differences between Mondays and Fridays would be seen in both groups. Interpretation is much harder if we have no controls.

2. EasyPeasy *is an intervention for preschoolers which "provides game ideas to the parents of preschool children to encourage play-based learning at home, with the aim of developing children's language development and self-regulation." In a report by the Education Endowment Foundation, the rationale is given as follows: "The assumption within the theory of change is that the EasyPeasy intervention will change child self-regulation which will lead to accelerated development in language and communication and improved school readiness. The expectation is that this will be achieved through the nursery teachers engaging with the parents regarding EasyPeasy and the parents engaging with their children through the EasyPeasy games. As well as improved self-regulation and language and communication development from playing the games, the expectation is that there will also be an improved home learning environment due to greater parent-child interaction. The expected impact was that this will lead to an improvement in children's readiness to learn along with improved parental engagement with the school."*

Suppose you had funds to evaluate EasyPeasy. What would you need to measure to assess outcomes and mechanism? What would you conclude if the outcomes improved but the mechanism measures showed no change?

Comment: The theory of change is quite complex as it involves both aspects of child development and of the environment. Within the child, the ultimate outcome of interest appears to be language and communication skills, whereas self-regulation is a mechanism through which these skills are improved. In addition, to fully test the model, we would need measures of the engagement of nursery teachers with parents, the home learning environment, and parental engagement with school, all of which appear to be regarded as mechanisms that will contribute to child language development.

One point to note is that although the theory of change would predict that the amount of improvement in language will vary with the amount of improvement in self-regulation and environmental measures, if this is not found, we would need to consider whether the measures were sensitive and reliable enough to detect change – see Chapter 3.

Chapter 9

1. *A study by Imhof et al. (2023) evaluated a video-coaching intervention with parents of children involved in a Head Start Program in the USA. Read the methods of the study and consider what measures were taken to encourage families to participate. How successful were they? What impact is there on the study of: (a) failure to recruit sufficient families into the study, (b) drop-out by those who had been enrolled in the intervention?*

Comment: Failure to recruit sufficient families will affect the statistical power of the study, which is a topic covered by Chapter 13. In brief, it means that the study may fail to detect genuine effects of the intervention unless they are large – subtle improvements might be missed. As discussed in the current chapter, missing data from drop-outs can be handled statistically, provided the data meet certain assumptions. In this study, the researchers did a statistical test to demonstrate that the data could be regarded as "Missing Completely At Random". They did not use imputation to handle the missing data, but used a method called Full Information Maximum Likelihood estimation, which is briefly described here: https://www.missingdata.nl/missing-data/missing-data-methods/full-information-maximum-likelihood/.

Chapter 12

1. *Figure 12.3 shows data from a small intervention trial in the form of a boxplot. Check your understanding of the annotation in the top left of the plot. How can you interpret the p-value? What does the 95% CI refer to? Is the "mean.diff" value the same as shown in the plot?*

   Comment: To take the last question first, the "mean. diff" value of 9.11 is in fact slightly larger than the difference between the horizontal bars on the boxplot. This is because boxplots typically show the median rather than the mean group value. There are a great many ways of representing summary information, and the boxplot is one popular format. If you are unfamiliar with boxplots or any other types of graph you encounter, we recommend Googling for more information.

   The first line of the annotation gives the standard output for a t-test, including the t-value, degrees of freedom, and associated p-value.

The 95% confidence interval expresses the uncertainty around the estimate of the mean difference. Note that with such a small sample, you get a very large confidence interval. The confidence interval spans zero, reflecting the fact that the p-value is greater than .05 and so does not reach the conventional level of statistical significance.

Chapter 13

1. *We have two studies of a spelling intervention, both using increase in number of words correct on a spelling test as the outcome measure. In both studies, two groups of 20 children were compared. In study A, the intervention group gained 10 words on average, with standard deviation of 2, and the control group gained 5 words, also with standard deviation of 2. In study B, the intervention group gained 15 words on average, with standard deviation of 5, and the control group gained 8 words with standard deviation of 5. Which study provides stronger evidence of the effectiveness of the intervention?*

   Comment: Although study B leads to a larger average gain in words correct, (15-8) = 7, the standard deviation is high. Study A has a smaller absolute gain, but with a much smaller standard deviation. We can convert the results to effect sizes, which shows that for study A, Cohen's d is $(10\text{-}5)/2 = 2.5$, whereas in study B, Cohen's d is $(15\text{-}8)/5 = 1.4$. So the evidence is stronger for study A. Having said that, in this fictitious example, both effect sizes are very large. With real-life studies, we would typically get effect sizes in the range of .3 to .4.

Chapter 14

1. *Type the phrase "did not survive Bonferroni correction" and "intervention" into a search engine to find a study that corrected for multiple corrections. Was the Bonferroni correction applied appropriately? Compare the account of the results in the Abstract and the Results section. Was the Bonferroni correction taken seriously?*

   Comment: In terms of whether the Bonferroni correction was appropriate, you can check two things: first, was the alpha level adjusted appropriately for the number of tests? That is usually correct, as it is a simple sum! The other question to consider is whether the correction was applied to correlated variables. This is not an issue when the correction is applied to subgroups (e.g., males vs. females) but it can affect findings when there are multiple measures on individuals. As explained in Chapter 14, if a set of measures is strongly intercorrelated, then the Bonferroni correction may be over-conservative – i.e., it may lead to a type II error, where a true effect gets missed. However, it is not always possible to judge whether this is the case, as authors may not report the intercorrelations between measures. Finally, it is not uncommon to find authors apply a Bonferroni correction and then fail to take it seriously. This is often evident when the Abstract refers only to the uncorrected statistics.

Chapter 16

Here are some characteristics of mediators and moderators. See if you can work out whether M refers to mediation or moderation.

1. *In an M relationship, you can draw an arrow from intervention to M and then from M to the outcome.* **Mediation**

2. *M refers to the case when something acts upon the relationship between intevention and outcome, and changes its strength.* **Moderation**

3. *M can be thought of as a go-between for intervention and outcome: it is a variable that explains their relationship.* **Mediation**

4. *Figure 16.3 shows an intervention where both mediation and moderation are involved.*

    Comment: M1 is moderator and M2 is mediator

Chapter 19

1. *Re: Calder et al. (2021) study: What could you conclude if:*

There was no difference between Groups 1 and 2 on past tense items at phase 4, with neither group showing any improvement over baseline?

Comment: This would suggest the intervention was ineffective, at least over the time scale considered here.

There was no difference between Groups 1 and 2 at phase 4, with both groups showing significant improvement over baseline?

Comment: This would suggest that improvement in both groups might be due to practice on the test items.

Group 1 improved at phase 4, but Group 2 did not improve between phase 4 and phase 6?

Comment: This would be a puzzling result, that in effect is a failure of replication of the beneficial effect seen in Group 1. Failures to replicate can just reflect the play of chance – the improvement in Group 1 could be a fluke – or it could indicate that there are individual differences in response to intervention, and by chance, those in group 2 contained more cases who were unresponsive.

Is it realistic to consider a possible "nocebo" effect during the extended waitlist for Group 2? Is there any way to check for this?

Comment: It is possible that children may get bored at repeatedly being assessed, especially if they are not receiving any help to master the test items. If so, then their performance might actually deteriorate across the baseline period.

Chapter 20

1. *Re: Koutsoftas study: Consider the following questions about this study.*

   a. *What kind of design is this?*

      Comment: Multiple baseline across participants: comparison between 2 groups who are initially observed over a baseline period, and then have intervention introduced at different time-points.

   b. *How well does this design guard against the biases shown in the first 5 rows of Table 20.3?*

      Comment: Inclusion of a baseline period establishes how much change we might see simply because of spontaneous improvement, practice effects, and regression to the mean, and also helps establish how noisy the dependent variable is.

   c. *Could the fact that intervention was delivered in small groups affect study validity? (Clue: see Chapter 18).*

      Comment: This does make this more like a clustered trial, and introduces the possibility that those within a group might influence one another.

d. *If you were designing a study to follow up on this result, what changes might you make to the study design?*

Comment: It might be useful to have more baseline observations, to apply the intervention to individuals rather than groups, and/or to include another 'control' measure tapping into skills that were not trained.

e. *What would be the logistic challenges in implementing these changes?*

Comment: The methodological changes suggested in (d) would make the study harder to do; having more sessions could tax the patience of children and teachers, and might be difficult to accommodate with the school timetable; similarly, applying the intervention to individuals would require more personnel. It might also change the nature of the intervention if interactions between children are important. Adding another control measure would lengthen the amount of assessment, which could be difficult with young children.

3. *Re: Calder et al: Once you have studied Figure 20.7, consider whether you think the inclusion of the probes has strengthened your confidence in the conclusion that the intervention is effective.*

Comment: The intervention effect was more pronounced for the trained items, and so cannot simply be attributed to practice or spontaneous improvement. This study illustrates how inclusion of control items that were not trained can add converging evidence to support the conclusion that an intervention is effective.

Chapter 24

1. *Read this preregistration of a study on Open Science Framework: https://osf.io/ecswy, and compare it with the article reporting results here: https://link.springer.com/article/10.1007/s11121-022-01455-4. Note any points where the report of the study diverges from the preregistration, and consider why this might have happened. Do the changes from preregistration influence the conclusions you can draw from the study?*

Comment: Compare your analysis with the evaluation of the study that was posted on Pubpeer here: https://pubpeer.com/publications/1AB6D8120F565 F8660D2933A87AB87.

References

Aksayli, N. D., Sala, G., & Gobet, F. (2019). The cognitive and academic benefits of Cogmed: A meta-analysis. *Educational Research Review, 27*, 229–243. https://doi.org/10.1016/j.edurev.2019.04.003

Armson, J., & Stuart, A. (1998). Effect of extended exposure to frequency-altered feedback on stuttering during reading and monologue. *Journal of Speech, Language, and Hearing Research, 41*(3), 479–490. https://doi.org/10.1044/jslhr.4103.479

Barber, T. X., & Silver, M. J. (1968). Fact, fiction, and the experimenter bias effect. *Psychological Bulletin, 70*(6, Pt.2), 1–29. https://doi.org/10.1037/h0026724

Bem, D. J. (2004). Writing the empirical journal article. In J. M. Darley, H. L. Roediger III, & M. P. Zanna (Eds.), *The Compleat Academic: A Career Guide, 2nd ed* (pp. 185–219). American Psychological Association.

Benjamin, D. J., Berger, J. O., Johannesson, M., Nosek, B. A., Wagenmakers, E.-J., Berk, R., Bollen, K. A., Brembs, B., Brown, L., Camerer, C., Cesarini, D., Chambers, C. D., Clyde, M., Cook, T. D., De Boeck, P., Dienes, Z., Dreber, A., Easwaran, K., Efferson, C., ... Johnson, V. E. (2018). Redefine statistical significance. *Nature Human Behaviour, 2*(1), 6–10. https://doi.org/10.1038/s41562-017-0189-z

Best, W., Greenwood, A., Grassly, J., Herbert, R., Hickin, J., & Howard, D. (2013). Aphasia rehabilitation: Does generalisation from anomia therapy occur and is it predictable? A case series study. *Cortex, 49*(9), 2345–2357. https://doi.org/10.1016/j.cortex.2013.01.005

Bishop, D. V. M. (2013a). Good and bad news on the phonics screen. In *BishopBlog*. http://deevybee.blogspot.com/2013/10/good-and-bad-news-on-phonics-screen.html

Bishop, D. V. M. (2013b). High-impact journals: Where newsworthiness trumps methodology. In *BishopBlog*. http://deevybee.blogspot.com/2013/03/high-impact-journals-where.html

Bishop, D. V. M. (2013c). Research Review: Emanuel Miller Memorial Lecture 2012 – Neuroscientific studies of intervention for language impairment in children: Interpretive and methodological problems. *Journal of Child Psychology and Psychiatry, and Allied Disciplines, 54*(3), 247–259. https://doi.org/10.1111/jcpp.12034

Bishop, D. V. M. (2019). Rein in the four horsemen of irreproducibility. *Nature, 568*(7753), 435–435. https://doi.org/10.1038/d41586-019-01307-2

Bishop, D. V. M. (2020). The psychology of experimental psychologists: Overcoming cognitive constraints to improve research: The 47th Sir Frederic Bartlett Lecture. *Quarterly Journal of Experimental Psychology (2006)*, *73*(1), 1–19. https://doi.org/10.1177/1747021819886519

Bishop, D. V. M. (2023). Using multiple outcomes in intervention studies: Improving power while controlling type I errors [version 2; peer review: 2 approved]. *F1000Research*, *10*, 991. https://doi.org/10.12688/f1000research.73520.2

Bishop, D. V. M., Adams, C. V., & Rosen, S. (2006). Resistance of grammatical impairment to computerized comprehension training in children with specific and non-specific language impairments. *International Journal of Language & Communication Disorders*, *41*(1), 19–40. https://doi.org/10.1080/13682820500144000

Bishop, D. V. M., & Edmundson, A. (1987). Language-impaired 4-year-olds: Distinguishing transient from persistent impairment. *Journal of Speech and Hearing Disorders*, *52*(2), 156–173. https://doi.org/10.1044/jshd.5202.156

Bishop, D. V. M., Snowling, M. J., Thompson, P. A., & Greenhalgh, T. (2017). Phase 2 of CATALISE: A multinational and multidisciplinary Delphi consensus study of problems with language development: Terminology. *Journal of Child Psychology and Psychiatry*, *58*(10), 1068–1080. https://doi.org/10.1111/jcpp.12721

Bishop, D. V. M., Thompson, J., & Parker, A. J. (2022). Can we shift belief in the "Law of Small Numbers"? *Royal Society Open Science*, *9*(3), 211028. https://doi.org/10.1098/rsos.211028

Bishop, D. V. M., & Thompson, P. A. (2016). Problems in using p-curve analysis and text-mining to detect rate of p-hacking and evidential value. *PeerJ*, *4*, e1715. https://doi.org/10.7717/peerj.1715

Block, S., Ingham, R. J., & Bench, R. J. (1996). The effects of the Edinburgh masker on stuttering. *Australian Journal of Human Communication Disorders*, *24*(1), 11–18. https://doi.org/10.3109/asl2.1996.24.issue-1.02

Borenstein, M., Hedges, L., Higgins, J. P. T., & Rothstein, H. (2009). *Introduction to Meta-Analysis*. John Wiley & Sons, Ltd. https://doi.org/10.1002/9780470743386.fmatter

Bowen, C. (2020). Independent research and the Arrowsmith Program. *Perspectives on Language and Literacy*, *46*(1), 47–54.

Bricker, D., & Squires, J. (1999). *Ages and Stages Questionnaires: A Parent-Completed, Child-Monitoring System, 2nd ed.* Paul H. Brookes.

Brown, R. (1973). *A First Language*. Harvard University Press.

Bull, L. (2007). Sunflower therapy for children with specific learning difficulties (dyslexia): A randomised, controlled trial. *Complementary Therapies in Clinical Practice*, *13*(1), 15–24. https://doi.org/10.1016/j.ctcp.2006.07.003

Burgoyne, K., Duff, F. J., Clarke, P. J., Buckley, S., Snowling, M. J., & Hulme, C. (2012). Efficacy of a reading and language intervention for children with Down syndrome: A randomized controlled trial. *Journal of*

Child Psychology and Psychiatry, and Allied Disciplines, 53(10), 1044–1053. https://doi.org/10.1111/j.1469-7610.2012.02557.x

Burgoyne, K., Duff, F. J., Clarke, P. J., Buckley, S., Snowling, M. J., & Hulme, C. (2016). *Reading and language intervention for children with Down syndrome: Experimental data [data collection], UK data service. SN: 852291.* https://doi.org/10.5255/UKDA-SN-852291

Burgoyne, K., Gardner, R., Whiteley, H., Snowling, M. J., & Hulme, C. (2018). Evaluation of a parent-delivered early language enrichment programme: Evidence from a randomised controlled trial. *Journal of Child Psychology and Psychiatry, and Allied Disciplines, 59*(5), 545–555. https://doi.org/10.1111/jcpp.12819

Button, K. (2018). Reboot undergraduate courses for reproducibility. *Nature, 561*(7723), 287–287. https://doi.org/10.1038/d41586-018-06692-8

Button, K. (2020). Supporting "team science." *The Psychologist, 33,* 30–33.

Calder, S. D., Claessen, M., Ebbels, S., & Leitão, S. (2021). The efficacy of an explicit intervention approach to improve past tense marking for early school-age children with Developmental Language Disorder. *Journal of Speech, Language, and Hearing Research, 64*(1), 91–104. https://doi.org/10.1044/2020_JSLHR-20-00132

Campbell, M., & Walters, S. (2014). *How to Design, Analyse and Report Cluster Randomised Trials in Medicine and Health Related Research.* Wiley.

Centre, N. C. B. R. (n.d.). Statistical assessment of reliability and validity. In *DAPA Measurement Toolkit.* https://www.measurement-toolkit.org/concepts/statistical-assessment

Chalmers, I., Bracken, M. B., Djulbegovic, B., Garattini, S., Grant, J., Gülme-zoglu, A. M., Howells, D. W., Ioannidis, J. P. A., & Oliver, S. (2014). How to increase value and reduce waste when research priorities are set. *The Lancet, 383*(9912), 156–165. https://doi.org/10.1016/S0140-6736(13)62229-1

Cheverud, J. M. (2001). A simple correction for multiple comparisons in interval mapping genome scans. *Heredity, 87*(1), 52–58. https://doi.org/10.1046/j.1365-2540.2001.00901.x

Corder, G. W., & Foreman, D. I. (2014). *Nonparametric Statistics: A Step-by-Step Approach.* Wiley.

Crystal, D., Fletcher, P., & Garman, M. (1977). *The Grammatical Analysis of Language Disability: A Procedure for Assessment and Remediation.* Edward Arnold.

Cunningham, J. A., Kypri, K., & McCambridge, J. (2013). Exploratory randomized controlled trial evaluating the impact of a waiting list control design. *BMC Medical Research Methodology, 13*(1), 150. https://doi.org/10.1186/1471-2288-13-150

De Angelis, C., Drazen, J. M., Frizelle, F. A., Haug, C., Hoey, J., Horton, R., Kotzin, S., Laine, C., Marusic, A., Overbeke, A. J. P. M., Schroeder, T. V., Sox, H. C., Van Der Weyden, M. B., & International Committee of Medical Journal Editors. (2004). Clinical trial registration: A statement from the International Committee of Medical Journal Editors. *Lancet*

(London, England), 364 (9438), 911–912. https://doi.org/10.1016/S0140-6736(04)17034-7

De Vries, Y. A., Roest, A. M., Jonge, P. de, Cuijpers, P., Munafò, M. R., & Bastiaansen, J. A. (2018). The cumulative effect of reporting and citation biases on the apparent efficacy of treatments: The case of depression. *Psychological Medicine, 48* (15), 2453–2455. https://doi.org/10.1017/S0033291718001873

Denman, D., Speyer, R., Munro, N., Pearce, W. M., Chen, Y.-W., & Cordier, R. (2017). Psychometric properties of language assessments for children aged 4–12 years: A systematic review. *Frontiers in Psychology, 8.* https://doi.org/10.3389/fpsyg.2017.01515

Derringer, J. (2018). *A simple correction for non-independent tests.* PsyArXiv. https://doi.org/10.31234/osf.io/f2tyw

Dipper, L., Marshall, J., Boyle, M., Botting, N., Hersh, D., Pritchard, M., & Cruice, M. (2020). Treatment for improving discourse in aphasia: A systematic review and synthesis of the evidence base. *Aphasiology.* https://doi.org/10.1080/02687038.2020.1765305

Duff, F. J., Mengoni, S. E., Bailey, A. M., & Snowling, M. J. (2015). Validity and sensitivity of the phonics screening check: Implications for practice. *Journal of Research in Reading, 38* (2), 109–123. https://doi.org/10.1111/1467-9817.12029

Ebbels, S. H., McCartney, E., Slonims, V., Dockrell, J. E., & Norbury, C. F. (2019). Evidence-based pathways to intervention for children with language disorders. *International Journal of Language & Communication Disorders, 54* (1), 3–19. https://doi.org/10.1111/1460-6984.12387

Eekhout, I. (2023). Don't miss out! In *Iris Eekhout | Missing data.* http://www.missingdata.nl/.

Elliott, S. A., & Brown, J. S. L. (2002). What are we doing to waiting list controls? *Behaviour Research and Therapy, 40* (9), 1047–1052. https://doi.org/10.1016/s0005-7967(01)00082-1

Epskamp, S. (2018). StatcheckTheGame. In *StatcheckTheGame.* https://sachaepskamp.github.io/StatcheckTheGame/.

Ferguson, C. J., & Heene, M. (2012). A vast graveyard of undead theories: Publication bias and psychological science's aversion to the null. *Perspectives on Psychological Science, 7* (6), 555–561. https://doi.org/10.1177/1745691612459059

Fielding, H. (1996). *Bridget Jones' Diary.* Picador.

Forsythe, L. P., Carman, K. L., Szydlowski, V., Fayish, L., Davidson, L., Hickam, D. H., Hall, C., Bhat, G., Neu, D., Stewart, L., Jalowsky, M., Aronson, N., & Anyanwu, C. U. (2019). Patient engagement in research: Early findings from the patient-centered outcomes research institute. *Health Affairs, 38* (3), 359–367. https://doi.org/10.1377/hlthaff.2018.05067

Freedman, B. (1987). Equipoise and the ethics of clinical research. *New England Journal of Medicine, 317* (3), 141–145. https://doi.org/10.1056/NEJM198707163170304

Friedman, L. S., & Richter, E. D. (2004). Relationship between conflicts of interest and research results. *Journal of General Internal Medicine, 19*(1), 51–56. https://doi.org/10.1111/j.1525-1497.2004.30617.x

Frizelle, P., Thompson, P., Duta, M., & Bishop, D. V. M. (2019). Assessing children's understanding of complex syntax: A comparison of two methods. *Language Learning, 69*(2), 255–291. https://doi.org/10.1111/lang.12332

Fugard, A. (2023). Regression to the mean – Andi Fugard. In *Personal website.* https://www.andifugard.info/regression-to-the-mean/

Furukawa, T. A., Noma, H., Caldwell, D. M., Honyashiki, M., Shinohara, K., Imai, H., Chen, P., Hunot, V., & Churchill, R. (2014). Waiting list may be a nocebo condition in psychotherapy trials: A contribution from network meta-analysis. *Acta Psychiatrica Scandinavica, 130*(3), 181–192. https://doi.org/10.1111/acps.12275

Garralda, E., Dienstmann, R., Piris-Giménez, A., Braña, I., Rodon, J., & Tabernero, J. (2019). New clinical trial designs in the era of precision medicine. *Molecular Oncology, 13*(3), 549–557. https://doi.org/10.1002/1878-0261.12465

Gelman, A., & Hill, J. (2007). *Data Analysis Using Regression and Multi-level/Hierarchical Models (Analytical Methods for Social Research).* Cambridge University Press.

Gillam, R. B., Loeb, D. F., Hoffman, L. M., Bohman, T., Champlin, C. A., Thibodeau, L., Widen, J., Brandel, J., & Friel-Patti, S. (2008). The efficacy of Fast ForWord-Language intervention in school-age children with language impairment: A randomized controlled trial. *Journal of Speech, Language, and Hearing Research, 51*(1), 97–119. https://doi.org/10.1044/1092-4388(2008/007)

Goldacre, B., Drysdale, H., Dale, A., Milosevic, I., Slade, E., Hartley, P., Marston, C., Powell-Smith, A., Heneghan, C., & Mahtani, K. R. (2019). COMPare: A prospective cohort study correcting and monitoring 58 misreported trials in real time. *Trials, 20*(1), 118. https://doi.org/10.1186/s13063-019-3173-2

Graham, P. W., Kim, M. M., Clinton-Sherrod, A. M., Yaros, A., Richmond, A. N., Jackson, M., & Corbie-Smith, G. (2016). What is the role of culture, diversity, and community engagement in transdisciplinary translational science? *Translational Behavioral Medicine, 6*(1), 115–124. https://doi.org/10.1007/s13142-015-0368-2

Greenhalgh, T., & Taylor, R. (1997). Papers that go beyond numbers (qualitative research). *BMJ (Clinical Research Ed.), 315*(7110), 740–743. https://doi.org/10.1136/bmj.315.7110.740

Greenwald, A. G. (1975). Consequences of prejudice against the null hypothesis. *Psychological Bulletin, 82*(1), 1–20. https://doi.org/10.1037/h0076157

Hardwicke, T. E., & Wagenmakers, E.-J. (2021). *Preregistration: A pragmatic tool to reduce bias and calibrate confidence in scientific research.* MetaArXiv. https://doi.org/10.31222/osf.io/d7bcu

Hemingway, P., & Brereton, N. J. (2009). *What is a systematic review? 2nd edition (online report).* http://www.bandolier.org.uk/painres/download/whatis/Syst-review.pdf

Henrich, J., Heine, S. J., & Norenzayan, A. (2010). Most people are not WEIRD. *Nature, 466*(7302), 29–29. https://doi.org/10.1038/466029a

Hernan, M. (2018). Causal inference from observational data. In *Miguel Hernan's Faculty Website.* https://www.hsph.harvard.edu/miguel-hernan/research/causal-inference-from-observational-data/.

Hey, S. P., & Kimmelman, J. (2014). The questionable use of unequal allocation in confirmatory trials. *Neurology, 82*(1), 77–79. https://doi.org/10.1212/01.wnl.0000438226.10353.1c

Higgins, J. P. T., Thomas, J., Chandler, J., Cumpston, M., Li, T., Page, M. J., & Welch, V. A. (Cochrane, 2021). *Cochrane Handbook for Systematic Reviews of Interventions, version 6.2 (updated February 2021).* https://training.cochrane.org/handbook/current.

Hoare, Z. S., Whitaker, C. J., & Whitaker, R. (2013). Introduction to a generalized method for adaptive randomization in trials. *Trials, 14*(1), 19. https://doi.org/10.1186/1745-6215-14-19

Holman, L., Head, M. L., Lanfear, R., & Jennions, M. D. (2015). Evidence of experimental bias in the life sciences: Why we need blind data recording. *PLOS Biology, 13*(7), e1002190. https://doi.org/10.1371/journal.pbio.1002190

Horner, R. H., & Odom, S. L. (2014). Constructing single-case research designs: Logic and options. In T. R. Kratochwill & J. R. Levin (Eds.), *Single-case Intervention Research: Methodological and Statistical Advances* (pp. 27–51). American Psychological Association.

Huang, F. L. (2018). Using instrumental variable estimation to evaluate randomized experiments with imperfect compliance. *Practical Assessment, Research, and Evaluation, 23*(2), Available at: https://scholarworks.umass.edu/pare/vol23/iss1/2. https://doi.org/10.7275/k0p6-yj16

Hulme, C., Bowyer-Crane, C., Carroll, J. M., Duff, F. J., & Snowling, M. J. (2012). The causal role of phoneme awareness and letter-sound knowledge in learning to read: Combining intervention studies with mediation analyses. *Psychological Science, 23*(6), 572–577. https://doi.org/10.1177/0956797611435921

Huntington-Klein, N. (2023). *The Effect: An Introduction to Research Design and Causality.* CRC Press.

Imhof, A., Liu, S., Schlueter, L., Phu, T., Watamura, S., & Fisher, P. (2023). Improving children's expressive language and auditory comprehension through responsive caregiving: Evidence from a randomized controlled trial of a strength-based video-coaching intervention. *Prevention Science, 24*(1), 84–93. https://doi.org/10.1007/s11121-022-01455-4

Ismay, C., & Kim, A. Y. (2023). *ModernDive: Statistical inference via data science.* https://moderndive.com/.

Jussim, L., & Harber, K. D. (2005). Teacher expectations and self-fulfilling prophecies: Knowns and unknowns, resolved and unresolved controversies. *Personality and Social Psychology Review, 9*(2), 131–155. https://doi.org/10.1207/s15327957pspr0902_3

Kemp-Koo, D. (2013). *A case study of the Learning Disabilities Association of Saskatchewan (LDAS) Arrowsmith program* [Doctoral thesis, University of Saskatchewan]. https://harvest.usask.ca/handle/10388/ETD-2013-11-1268

Kerr, N. L. (1998). HARKing: Hypothesizing after the results are known. *Personality and Social Psychology Review, 2*(3), 196–217. https://doi.org/10.1207/s15327957pspr0203_4

Koutsoftas, A. D., Harmon, M. T., & Gray, S. (2009). The effect of Tier 2 intervention for phonemic awareness in a response-to-intervention model in low-income preschool classrooms. *Language, Speech, and Hearing Services in Schools, 40*(2), 116–130. https://doi.org/10.1044/0161-1461(2008/07-0101)

Kraemer, H. C., Wilson, G. T., Fairburn, C. G., & Agras, W. S. (2002). Mediators and moderators of treatment effects in randomized clinical trials. *Archives of General Psychiatry, 59*(10), 877–883. https://doi.org/10.1001/archpsyc.59.10.877

Kraft, M. A. (2023). The effect-size benchmark that matters most: Education interventions often fail. *Educational Researcher, 52*(3), 183–187.

Krasny-Pacini, A., & Evans, J. (2018). Single-case experimental designs to assess intervention effectiveness in rehabilitation: A practical guide. *Annals of Physical and Rehabilitation Medicine, 61*(3), 164–179. https://doi.org/10.1016/j.rehab.2017.12.002

Kratochwill, T. R., & Levin, J. R. (Eds.). (2014). *Single-case Intervention Research: Methodological and Statistical Advances* (pp. xiii, 366). American Psychological Association. https://doi.org/10.1037/14376-000

Kratochwill, T. R., Levin, J. R., Horner, R. H., & Swoboda, C. M. (2014). Visual analysis of single-case intervention research: Conceptual and methodological issues. In T. R. Kratochwill & J. R. Levin (Eds.), *Single-case Intervention Research: Methodological and Statistical Advances* (pp. 91–125). American Psychological Association. https://doi.org/10.1037/14376-004

Lajeunesse, M. J. (2016). Facilitating systematic reviews, data extraction, and meta-analysis with the metagear package for r. *Methods in Ecology and Evolution, 7*, 323–330.

Lakens, D. (2021). *Sample Size Justification.* PsyArXiv. https://doi.org/10.31234/osf.io/9d3yf

Lakens, D., Adolfi, F. G., Albers, C. J., Anvari, F., Apps, M. A. J., Argamon, S. E., Baguley, T., Becker, R. B., Benning, S. D., Bradford, D. E., Buchanan, E. M., Caldwell, A. R., Van Calster, B., Carlsson, R., Chen, S.-C., Chung, B., Colling, L. J., Collins, G. S., Crook, Z., . . . Zwaan, R. A. (2018). Justify your alpha. *Nature Human Behaviour, 2*(3), 168–171. https://doi.org/10.1038/s41562-018-0311-x

Law, J., Garrett, Z., & Nye, C. (2003). Speech and language therapy interventions for children with primary speech and language delay or disorder.

Cochrane Database of Systematic Reviews, 3. https://doi.org/10.1002/1465 1858.CD004110

Ledford, J. R., Barton, E. E., Severini, K. E., & Zimmerman, K. N. (2019). A primer on single-case research designs: Contemporary use and analysis. *American Journal on Intellectual and Developmental Disabilities, 124*(1), 35. https://doi.org/10.1352/1944-7558-124.1.35

Leng, G., & Leng, R. I. (2020). *The Matter of Facts: Skepticism, Persuasion, and Evidence in Science.* MIT Press.

Leniston, H., & Ebbels, S. (2021). Investigation into the effectiveness of electropalatography in treating persisting speech sound disorders in adolescents with co-occurring developmental language disorder. *Clinical Linguistics & Phonetics,* 1–16. https://doi.org/10.1080/02699206.2021.1957022

Levitt, S. D., & List, J. A. (2011). Was there really a Hawthorne Effect at the Hawthorne Plant? An analysis of the original illumination experiments. *American Economic Journal: Applied Economics, 3*(1), 224–238. https://doi.org/10.1257/app.3.1.224

Lo, B., & Field, M. J. (2009). *Conflict of Interest in Medical Research, Education, and Practice.* National Academies Press.

Loeb, D. F., Stoke, C., & Fey, M. E. (2001). Language changes associated with Fast ForWord-language: Evidence from case studies. *American Journal of Speech-Language Pathology, 10*(3), 216–230. https://doi.org/10.1044/1058-0360(2001/020)

Lord, F. M., Novick, M. R., & Birnbaum, A. (1968). *Stastistical Theories of Mental Test Scores.* Addison-Wesley.

Ludlow, C. (2013). Need for adaptive research designs in speech-language pathology. In *ASHA Journals Academy.* https://academy.pubs. asha.org/2013/11/need-for-adaptive-research-designs-in-speech-language-pathology/.

Macnamara, B. N., & Burgoyne, A. P. (2023). Do growth mindset interventions impact students' academic achievement? A systematic review and meta-analysis with recommendations for best practices. *Psychological Bulletin, 149*(3-4), 133–173. https://doi.org/10.1037/bul0000352

Mahoney, M. J. (1976). *Scientist as Subject: The Psychological Imperative.* Ballinger Publishing Company.

McGillion, M., Pine, J. M., Herbert, J. S., & Matthews, D. (2017). A randomised controlled trial to test the effect of promoting caregiver contingent talk on language development in infants from diverse socioeconomic status backgrounds. *Journal of Child Psychology and Psychiatry, 58*(10), 1122–1131. https://doi.org/10.1111/jcpp.12725

Mirza, R., Punja, S., Vohra, S., & Guyatt, G. (2017). The history and development of N-of-1 trials. *Journal of the Royal Society of Medicine, 110*(8), 330–340. https://doi.org/10.1177/0141076817721131

Moher, D., Hopewell, S., Schulz, K. F., Montori, V., Gøtzsche, P. C., Devereaux, P. J., Elbourne, D., Egger, M., & Altman, D. G. (2010). CONSORT 2010 explanation and elaboration: Updated guidelines for reporting parallel group randomised trials. *British Medical Journal (Clinical Research Ed.)*, *340*, c869. https://doi.org/10.1136/bmj.c869

Morris, D., Fraser, S., & Wormald, R. (2007). Masking is better than blinding. *British Medical Journal*, *334*(7597), 799–799. https://doi.org/10.1136/bmj.39175.503299.94

Morrow, E. L., Duff, M. C., & Mayberry, L. S. (2022). Mediators, moderators, and covariates: Matching analysis approach for improved precision in cognitive-communication rehabilitation research. *Journal of Speech, Language, and Hearing Research*, *65*(11), 4159–4171. https://doi.org/10.1044/2022_JSLHR-21-00551

O'Connell, N. S., Dai, L., Jiang, Y., Speiser, J. L., Ward, R., Wei, W., Carroll, R., & Gebregziabher, M. (2017). Methods for analysis of pre-post data in clinical research: A comparison of five common methods. *Journal of Biometrics & Biostatistics*, *8*(1), 1–8. https://doi.org/10.4172/2155-6180.1000334

Pallmann, P., Bedding, A. W., Choodari-Oskooei, B., Dimairo, M., Flight, L., Hampson, L. V., Holmes, J., Mander, A. P., Odondi, L., Sydes, M. R., Villar, S. S., Wason, J. M. S., Weir, C. J., Wheeler, G. M., Yap, C., & Jaki, T. (2018). Adaptive designs in clinical trials: Why use them, and how to run and report them. *BMC Medicine*, *16*(1), 29. https://doi.org/10.1186/s12916-018-1017-7

Parker, R. I., Vannest, K. J., & Davis, J. L. (2014). Non-overlap analysis for single-case research. In T. R. Kratochwill & J. R. Levin (Eds.), *Single-case Intervention Research: Methodological and Statistical Advances* (pp. 127–151). American Psychological Association. https://doi.org/10.1037/14376-005

Patsopoulos, N. A. (2011). A pragmatic view on pragmatic trials. *Dialogues in Clinical Neuroscience*, *13*(2), 217–224.

Perdices, M., & Tate, R. L. (2009). Single-subject designs as a tool for evidence-based clinical practice: Are they unrecognised and undervalued? *Neuropsychological Rehabilitation*, *19*(6), 904–927. https://doi.org/10.1080/09602010903040691

Phillips, N. (2017). *Yarrr: A companion to the e-book "YaRrr!: The pirate's guide to r"*. https://CRAN.R-project.org/package=yarrr

Pocock, S. J., & Simon, R. (1975). Sequential treatment assignment with balancing for prognostic factors in the controlled clinical trial. *Biometrics*, *31*(1), 103–115.

Poldrack, R. (2018). *Statistical thinking for the 21st century: An open source textbook for statistics, with companions for R and Python*. https://github.com/statsthinking21/statsthinking21.

Porzsolt, F., Rocha, N. G., Toledo-Arruda, A. C., Thomaz, T. G., Moraes, C., Bessa-Guerra, T. R., Leão, M., Migowski, A., Araujo da Silva, A. R., & Weiss, C. (2015). Efficacy and effectiveness trials have different goals, use different tools, and generate different messages. *Pragmatic and Observational Research*, *6*, 47–54. https://doi.org/10.2147/POR.S89946

Pustejovsky, J. E. (2018). Easily simulate thousands of single-case designs. In *James E. Pustejovsky*. https://www.jepusto.com/easily-simulate-thousands-of-single-case-designs/.

Quintana, D. S., & Williams, D. R. (2018). Bayesian alternatives for common null-hypothesis significance tests in psychiatry: A non-technical guide using JASP. *BMC Psychiatry, 18*(1), 178. https://doi.org/10.1186/s12888-018-1761-4

Rabbitt, P., Diggle, P., Holland, F., & McInnes, L. (2004). Practice and drop-out effects during a 17-year longitudinal study of cognitive aging. *The Journals of Gerontology. Series B, Psychological Sciences and Social Sciences, 59*(2), P84–97. https://doi.org/10.1093/geronb/59.2.p84

Ratner, N. B., & MacWhinney, B. (2016). Your laptop to the rescue: Using the Child Language Data Exchange System Archive and CLAN Utilities to improve child language sample analysis. *Seminars in Speech and Language, 37*(2), 74–84. https://doi.org/10.1055/s-0036-1580742

Reise, S. P., Ainsworth, A. T., & Haviland, M. G. (2005). Item Response Theory: Fundamentals, applications, and promise in psychological research. *Current Directions in Psychological Science, 14*(2), 95–101. https://doi.org/10.1111/j.0963-7214.2005.00342.x

Renfrew, C. (1967). *Action Picture Test*. Catherine Renfrew.

Renfrew, C. (2010). *Bus Story Test: Revised Edition*. Routledge.

Rindskopf, D. M., & Ferron, J. M. (2014). Using multilevel models to analyze single-case design data. In *Single-Case Intervention Research: Methodological and Statistical Advances* (pp. 221–246). American Psychological Association. https://doi.org/10.1037/14376-008

Rioux, C., & Little, T. D. (2021). Missing data treatments in intervention studies: What was, what is, and what should be. *International Journal of Behavioral Development, 45*(1), 51–58. https://doi.org/10.1177/0165025419880609

Robinson, K. A., & Goodman, S. N. (2011). A systematic examination of the citation of prior research in reports of randomized, controlled trials. *Annals of Internal Medicine, 154*(1), 50–55. https://doi.org/10.7326/0003-4819-154-1-201101040-00007

Rosenthal, R. (1979). The file drawer problem and tolerance for null results. *Psychological Bulletin, 86*(3), 638–641. https://doi.org/10.1037/0033-2909.86.3.638

Rosenthal, R., & Fode, K. L. (1963). The effect of experimenter bias on the performance of the albino rat. *Behavioral Science, 8*(3), 183–189. https://doi.org/10.1002/bs.3830080302

Rosenthal, R., & Jacobson, L. (1968). *Pygmalion in the Classroom: Teacher Expectations and Student Intellectual Development*. Holt.

Rothwell, P. M. (2005). External validity of randomised controlled trials: "To whom do the results of this trial apply?" *The Lancet, 365*(9453), 82–93. https://doi.org/10.1016/S0140-6736(04)17670-8

Schönbrodt, F. (2016). *Introducing the P-Hacker App: Train Your Expert p-hacking Skills.* https://www.nicebread.de/introducing-p-hacker/

Schulz, K. F., Altman, D. G., & Moher, D. (2010). CONSORT 2010 Statement: Updated guidelines for reporting parallel group randomised trials. *British Medical Journal, 340*, c332. https://doi.org/10.1136/bmj.c332

Scott, N. W., McPherson, G. C., Ramsay, C. R., & Campbell, M. K. (2002). The method of minimization for allocation to clinical trials. A review. *Controlled Clinical Trials, 23*(6), 662–674. https://doi.org/10.1016/s0197-2456(02)00242-8

Senn, S. (2018). Statistical pitfalls of personalized medicine. *Nature, 563*(7733), 619–621. https://doi.org/10.1038/d41586-018-07535-2

Sibbald, B., & Roberts, C. (1998). Understanding controlled trials: Crossover trials. *British Medical Journal, 316*(7146), 1719–1720.

Simmons, J. P., Nelson, L. D., & Simonsohn, U. (2011). False-positive psychology: Undisclosed flexibility in data collection and analysis allows presenting anything as significant. *Psychological Science, 22*(11), 1359–1366. https://doi.org/10.1177/0956797611417632

Simonsohn, U., Nelson, L. D., & Simmons, J. P. (2014). P-curve: A key to the file-drawer. *Journal of Experimental Psychology. General, 143*(2), 534–547. https://doi.org/10.1037/a0033242

Singal, A. G., Higgins, P. D. R., & Waljee, A. K. (2014). A primer on effectiveness and efficacy trials. *Clinical and Translational Gastroenterology, 5*(1), e45. https://doi.org/10.1038/ctg.2013.13

Steegen, S., Tuerlinckx, F., Gelman, A., & Vanpaemel, W. (2016). Increasing transparency through a multiverse analysis. *Perspectives on Psychological Science: A Journal of the Association for Psychological Science, 11*(5), 702–712. https://doi.org/10.1177/1745691616658637

Swain, N. R., Eadie, P. A., & Snow, P. C. (2020). Speech and language therapy for adolescents in youth justice: A series of empirical single-case studies. *International Journal of Language & Communication Disorders, 55*(4), 458–479. https://doi.org/10.1111/1460-6984.12529

Tate, R. L., Perdices, M., Rosenkoetter, U., McDonald, S., Togher, L., Shadish, W., Horner, R., Kratochwill, T., Barlow, D. H., Kazdin, A., Sampson, M., Shamseer, L., & Vohra, S. (2016). The Single-Case Reporting Guideline In BEhavioural Interventions (SCRIBE) 2016: Explanation and elaboration. *Archives of Scientific Psychology, 4*(1), 10. https://doi.org/10.1037/arc0000027

Tate, R. L., Perdices, M., Rosenkoetter, U., Shadish, W., Vohra, S., Barlow, D. H., Horner, R., Kazdin, A., Kratochwill, T., McDonald, S., Sampson, M., Shamseer, L., Togher, L., Albin, R., Backman, C., Douglas, J., Evans, J. J., Gast, D., Manolov, R., . . . Wilson, B. (2016). The Single-Case Reporting Guideline In Behavioural Interventions (SCRIBE) 2016 Statement. *Physical Therapy, 96*(7), e1–e10. https://doi.org/10.2522/ptj.2016.96.7.e1

Thorlund, K., Haggstrom, J., Park, J. J., & Mills, E. J. (2018). Key design considerations for adaptive clinical trials: A primer for clinicians. *British Medical Journal, 360*, k698. https://doi.org/10.1136/bmj.k698

Torgesen, J. K., Wagner, R. K., & Rashotte, C. A. (1999). *Test of Word Reading Efficiency*. Pro-Ed.

Treweek, S., & Zwarenstein, M. (2009). Making trials matter: Pragmatic and explanatory trials and the problem of applicability. *Trials, 10*(1), 37. https://doi.org/10.1186/1745-6215-10-37

Van Buuren, S. (2018). *Flexible Imputation of Missing Data, 2nd edition (online book: https://stefvanbuuren.name/fimd/)*. Chapman and Hall/CRC.

Varley, R., Cowell, P., Dyson, L., Inglis, L., Roper, A., & Whiteside, S. P. (2016). Self-administered computer therapy for apraxia of speech. *Stroke, 47*(3), 822–828. https://doi.org/10.1161/STROKEAHA.115.011939

Wason, J. M. S., Brocklehurst, P., & Yap, C. (2019). When to keep it simple – adaptive designs are not always useful. *BMC Medicine, 17*(1), 152. https://doi.org/10.1186/s12916-019-1391-9

Weiss, S. (1991). Stressors experienced by family caregivers of children with pervasive developmental disorders | SpringerLink. *Child Psychiatry and Human Development, 21*, 203–216.

Wiig, E. H., Secord, W., & Semel, E. (2006). *CELF-Preschool 2 UK*. Pearson Assessment.

Zhang, S., Heck, P. R., Meyer, M., Chabris, C. F., Goldstein, D. G., & Hofman, J. M. (2022). *An illusion of predictability in scientific results*. SocArXiv. https://doi.org/10.31235/osf.io/5tcgs

Zhang, X., & Tomblin, J. B. (2003). Explaining and controlling regression to the mean in longitudinal research designs. *Journal of Speech, Language, and Hearing Research, 46*(6), 1340–1351. https://doi.org/10.1044/1092-4388(2003/104)

Index

Printed in the United States
by Baker & Taylor Publisher Services